Thermal and Structural Electronic Packaging Analysis for Space and Extreme Environments

Taylor and Francis Series in Resilience and Sustainability in Civil, Mechanical, Aerospace and Manufacturing Engineering Systems

Series Editor
Mohammad Noori
Cal Poly San Luis Obispo

Published Titles

Resilience of Critical Infrastructure Systems
Emerging Developments and Future Challenges
Zhishen Wu, Xilin Lu, and Mohammad Noori

Experimental Vibration Analysis for Civil Structures
Testing, Sensing, Monitoring, and Control
Jian Zhang, Zhishen Wu, Mohammad Noori, and Yong Li

Reliability and Safety of Cable-Supported Bridges
Naiwei Lu, Yang Liu, and Mohammad Noori

Reliability-Based Analysis and Design of Structures and Infrastructure
Ehsan Noroozinejad Farsangi, Mohammad Noori, Paolo Gardoni, Izuru Takewaki, Humberto Varum, Aleksandra Bogdanovic

Seismic Analysis and Design using the Endurance Time Method
Edited by Homayoon E. Estekanchi and Hassan A. Vafai

Thermal and Structural Electronic Packaging Analysis for Space and Extreme Environments
Juan Cepeda-Rizo, Jeremiah Gayle, and Joshua Ravich

For more information about this series, please visit: www.routledge.com/Resilience-and-Sustainability-in-Civil-Mechanical-Aerospace-and-Manufacturing/book-series/ENG

Thermal and Structural Electronic Packaging Analysis for Space and Extreme Environments

Juan Cepeda-Rizo, Jeremiah Gayle, and Joshua Ravich

CRC Press
Taylor & Francis Group
Boca Raton London New York

CRC Press is an imprint of the
Taylor & Francis Group, an **informa** business

First edition published 2022
by CRC Press
6000 Broken Sound Parkway NW, Suite 300, Boca Raton, FL 33487–2742

and by CRC Press
2 Park Square, Milton Park, Abingdon, Oxon, OX14 4RN

Library of Congress Cataloging-in-Publication Data
Names: Cepeda-Rizo, Juan, author. | Gayle, Jeremiah, author. | Ravich, Joshua, author.
Title: Thermal and structural electronic packaging analysis for space and extreme environments / Juan Cepeda-Rizo, Jeremiah Gayle, and Joshua Ravich.
Description: First edition. | Boca Raton, FL : CRC Press, 2022. | Series: Resilience and sustainability in civil, mechanical, aerospace and manufacturing engineering systems | Includes bibliographical references and index.
Identifiers: LCCN 2021031791 (print) | LCCN 2021031792 (ebook) | ISBN 9781032160818 (hbk) | ISBN 9781032160856 (pbk) | ISBN 9781003247005 (ebk)
Subjects: LCSH: Electronic packaging. | Electronic apparatus and appliances—Reliability. | Space vehicles—Electronic equipment—Design and construction. | Extreme environments.
Classification: LCC TK7870.15 .C47 2022 (print) | LCC TK7870.15 (ebook) | DDC 621.381/046—dc23
LC record available at https://lccn.loc.gov/2021031791
LC ebook record available at https://lccn.loc.gov/2021031792

ISBN: 978-1-032-16081-8 (hbk)
ISBN: 978-1-032-16085-6 (pbk)
ISBN: 978-1-003-24700-5 (ebk)

DOI: 10.1201/9781003247005

Contents

Preface.. xi
Author Biographies.. xiii

Chapter 1   Introduction 1

Electronics in Space 1
The Importance of the Electronic Packaging Engineer...................... 1
Baked-in Reliability 5

Chapter 2   New Space.. 7

Focus on Reliability... 8
Focus on Repeatability 8
Focus on ROI (Return on Investment)... 8

Chapter 3   Thermal/Structural Challenges in Miniaturizing............................. 13

An Assessment of SpaceVNX by JPL Electronics Packaging Group 13

Chapter 4   Fundamentals of Heat Transfer by Conduction and Convection........ 19

Introduction 19
Conduction Heat Transfer 19
Convective Heat Transfer 20

Chapter 5   Fundamentals of Heat Transfer by Radiation.................................... 23

Introduction 23
Blackbody Radiation 24
Emissivity 26
Absorptivity.. 27
Reflectivity 29
Gray Surfaces 29
Diffuse Surfaces 30
Radiation Geometry 31

Chapter 6   The Multilayer Insulation (MLI) Blanket 33

Blanket Applications 33
Design (Driven by Requirements and Considerations) 34

Chapter 7   Heat Pipes... 41

    Variable Conductance Heat Pipes (VCHPs)................................ 41
    Cryogenic Switching Heat Pipes (CSHPs)................................. 42
    Water Sintered-Copper Heat Pipes.. 43
    Nano Heat Pipes (NHPs)... 43
    Testing 46
    I. Introduction.. 55
    II. Analysis of Heat Pipe Failure 56
    B. Wick Failure.. 59
    C. 1D Heat Transfer Models 61
    Testing 65
    Freeze/Thaw Test Results.. 68
    Conclusion.. 68
    Future Work.. 68
    Acknowledgement.. 68
    Phase Change γ-Alumina Aqueous-Based Nanofluid for
        Improving Heat Pipe Transient Efficiency (The Nano
        Heat Pipe) 70
    Summary and Conclusions... 77

Chapter 8   Convective Cooling of Semiconductors Using a Nanofluid.............. 79

    Introduction 79
    Forced Convection Flow.. 80
    Nanofluids for Cooling Electronics... 82
    Nanoscale Thermal Interface... 88
    Test Vehicle Design 88
    Experimental Setup 89
    Summary and Conclusions... 96
    Acknowledgement.. 98

Chapter 9   Power Systems: The Tesla Turbine... 99

    Introduction 99
    Arriving at a Closed-Form Solution 107
    Case Study, an Automotive Air Conditioner 110

Chapter 10   Electronics Design for Extreme Temperature and Pressure............ 123

Chapter 11   Characterization and Modeling of PWB Warpage and Its Effect
    on LGA Separable Interconnects 125

    Introduction 125
    Method of Investigation... 127

Results 131
Conclusion 138
Acknowledgements 139

Chapter 12 Resistor Networks 145

Conduction 145

Chapter 13 Thermal Analysis Case Studies 151

Introduction to the Case Studies 151
Subject: Thermal Analysis of Scanner Demodulator Electronics
    (SDE) for SHERLOC 151
Subject: SHERLOC Laser Power Supply Transient Thermal
    Analysis 157
Detailed Mode (100 Spectra) 159
U1—Thermal Strap Thermal Analysis at 55 °C Boundary
    Temperature 163
Subject: Universal System Transponder (UST) Thermal Analysis .. 163
Subject: Preliminary Power Switch Slice (PSS) Assembly
    Thermal Analysis 177

Chapter 14 Random Vibration Structural Analysis and Miles' Equation 191

Introduction 191
Limiting Factors 191
Load 191
Modal Response 192
Relative Displacement 192
Model 193
Acceleration Response 193
Relative Displacement and Spring Force 194
RMS and Standard Deviation 195

Chapter 15 Vibrational Analysis Case Studies 197

Introduction to the Vibrational Case Studies 197
Subject: Photodiode Board Assembly Structural Analysis 197
Subject: Sine Vibration Analysis of HRMR DDU for Sentinel-6 199
Subject: Laser Power Supply (LPS) Structural Analysis 208

Chapter 16 Creep Prediction of a Printed Wiring Board for Separable Land
            Grid Array Connector 227

Introduction 227
Mechanical Scheme 227

Prediction of Long-Term Creep 229
Experimental Setup 231
Results 232
Summary and Conclusion 233

Chapter 17 Operational Case Studies—Mars Surface Operations..................... 237

MSL Rear Hazcam Thermal Characterization 237
I. Introduction... 237
II. Description... 238
1. Engineering Cameras 238
2. Rear Hazcam Environment.. 240
3. Model Description.. 240
III. Testing of the Rear Hazcam on Mars 242
Hazcam Thermal Characterization and the ECAM Calculator 242
IV. Results 244
A. Hazcam Telemetry vs. ECAM Calculator 244
B. Model Prediction vs. Telemetry.................................... 245
V. Conclusion 245
VI. Future Work.. 245
VII. Acknowledgement 246

Laser Power Supply Thermo-Structural Analysis for the Mars
2020 Rover 247

Introduction 247
Design Consideration: Thermal and Electrical Isolation 247
Materials... 248
Thermal Results.. 248
Transient Thermal Analysis... 249
Steinberg Fatigue Analysis 249
Chassis Fasteners.. 251
Venting Analysis... 251
Conclusion.. 254
Acknowledgement... 258

Chapter 18 Operational Case Studies—Dawn Asteroid Mission....................... 259

Introduction 259
Thermal Design Based on Flight Proven Thermal Control
          Techniques... 259
Operation.. 263

Chapter 19 Standards 271

1.0 Structural Design and Test Requirements (NASA-
    STD-5001) [1] 271
2.0 Derating Standards ECSS-Q-ST-30–11-REV1 Derating EEE
    Components, Page 40 [2]... 272
3.0 GEVS NASA-STD-7000A[3] 275
4.0 SMC-160 [4] 283
5.0 Bolt Thermal Resistances—Spacecraft Thermal Control
    Handbook [5]... 286

Index 287

Preface

With the advent of CubeSats and miniature satellites, the space industry has seen a push in recent times towards commercialization and an ushering in of the New Space era. These small satellites have left the research and university stages and have seen more commercial and military applications and appeal. Start-ups are created daily, and venture capital has quickly seen this industry become highly lucrative, some predicting a trillion-dollar industry within the next 10 years. SpaceX has drastically reduced the cost of launch, and companies like Rocket Lab even offer monthly launches to put your satellite into orbit every month. Starlink is promising to disrupt the internet industry with its constellation of small satellites that will offer internet service to all corners of the world, while companies such as Planet Labs are offering real-time versions of the well-known Google Maps.

This race to space has quickly put reliability on the back burner. With smaller budgets than heavily government-funded juggernauts such as Boeing and Northrop Grumman, many of these startups cannot afford costly test-as-you-fly campaigns, making preemptive designing and computer analysis the more valuable. It is estimated that 50% of CubeSats and small satellites experience premature failure upon reaching orbit, with many never working at all. This book is written in a practical and useful manner to aid engineers in analyzing the harsh mechanical and thermal environments of space environments and to catch problems early in the design process before reaching the testing stage. The techniques are employed by NASA's Jet Propulsion Laboratory, who have been known to push the envelope across all aspects of space flight.

Author Biographies

Juan Cepeda-Rizo obtained his bachelor's degree in mechanical engineering and his master's in materials engineering from Cal Poly, San Luis Obispo in 1997. After graduation he worked for a semiconductor company doing electronic packaging analysis. He received his PhD from Claremont Graduate University in applied mathematics, and spent 13 years at NASA's Jet Propulsion Laboratory as a thermal systems engineer. He is currently working for Rocket Lab's Space Systems Group where his responsibilities include spacecraft thermal design and analysis and thermal-structural analysis of flight electronics.

Jeremiah Gayle obtained his bachelor's degree in mechanical engineering from Arizona State University and his master's degrees in systems engineering from Iowa State, and John Hopkins University in space systems engineering. He is pursuing a doctorate in aerospace engineering from Colorado State University and currently works for NASA's Jet Propulsion Laboratory Electronic Packaging Group where he conducts thermal and structural analysis of avionics.

Joshua Ravich is a supervisor for the Technology Infusion group at JPL where he works on the mechanical design and analysis of spacecraft systems. His recent work has included the Mars Helicopter project. He received a bachelor's degree in mechanical engineering from UC Berkeley and a master's in mechanical and aerospace engineering from the University of Michigan.

1 Introduction

ELECTRONICS IN SPACE

When we look at the constraints around building a spacecraft for space applications, whether they are for low earth orbit (LEO) or for visiting the outer planets, the power, thermal, and structural constraints of the design often boil down to keeping the electronics hardware on board safe, happy, and productive. When images from Mars come down and someone wonders why some are in black and white, the answer that we give at NASA is that color photos require more memory to store and more resources in general. Also, someone may ask, why is it that I can take high-resolution videos with my cell phone, in color, and pretty much anytime I want to and as many as I want, but the rover on Mars cannot do the same? The answer that I give is usually something like, "your cell phone doesn't have to survive brutal nighttime temperatures down to −100 °C, and during the day there is almost no air to aid in cooling." Worse yet, closer to home the moon has a scorching daytime surface temperature of +150 °C and a frigid nighttime surface temperature of −200 °C, all within the vacuum of space, and don't get me started on the extreme temperatures and pressures on the surface of Venus. All this, and the electronics also have to survive an extremely bumpy rocket launch that would destroy most laptop computers as well as all the tiny solder connections on your precious cell phone!

After working in the consumer electronics industry for 11 years, I decided to work for NASA's Jet Propulsion Laboratory as systems engineer. When I first looked at the electronics that were being sent to Mars aboard the Curiosity rover, I will never forget how my jaw dropped on how primitive they appeared. I mean Dual Inline Package (DIP) packages, and J-leaded transistors, and through-hole parts? Hot Air Solder Level (HASL) boards? Eutectic tin-lead solder? I thought I transported 20 years into the past. Where were the BGAs and all the flip chip components that I was so used to seeing? JPL is a legacy space company that tries to fly as many heritage parts as possible, and if that means flying DIP parts, then we will do that and have no qualms over it. All in the name of reliability, and that is why our Mars exploration rover, Opportunity, lasted for almost two decades on the surface of Mars.

The purpose of this book is not to set the clock back 20 years, but instead to apply the same principles of thermal and structural analysis [1] onto the world of commercial off-the-shelf (COTS) parts and demonstrate survivability of space and extreme environments.

THE IMPORTANCE OF THE ELECTRONIC PACKAGING ENGINEER

The electronic packaging (EP) engineer plays a key role in the design, analysis, and manufacturing of the electronics that will go to space [2]. No other engineering discipline requires such a broad knowledge base and covers many if not all the main engineering disciplines, including thermal, structural, materials, electronics, and

DOI: 10.1201/9781003247005-1

manufacturing. The goal of the EP engineer is to gather and combine components and assemblies in strict adherence to reducing size, weight, and power (SWaP) [2] yet still maintain reliability at the highest level. Traditionally, the packaging of electronics has been defined by two levels: Level 1—the components and parts (Figure 1.1), and Level 2—the subassemblies and top-level "electronic box" assembly (Figure 1.2). A third level (Level 3) is usually owned by the spacecraft mechanical and thermal designers and rarely involves the traditional EP engineer.

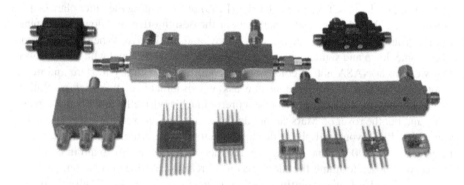

FIGURE 1.1 Level 1 includes flight-qualified components and parts.

(Source: www.rhombustechnologies.com.au/wp-content/uploads/2015/10/passive-components.png)

FIGURE 1.2 Level 2 electronics includes all the subassembly electronics board slices, connectors, and enclosures.

(Source: www.aeroespacial.sener/en/products/mtg-scan-assembly-electronics)

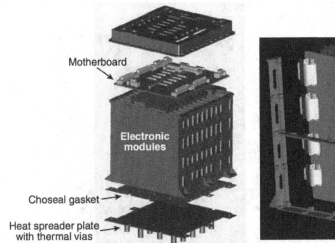

FIGURE 1.3 Typical "bread loaf" design consists of individual slices bolted together and heat conducted down the chassis rails to the heat spreader below [3].

Figure 1.3 shows the typical "bread loaf" chassis design where multiple slices are stacked together vertically and bolted to the radiator along the bottom attachment plate. This configuration allows closeness of the individual slice assemblies, which helps with speed of communication while offering a space savings on the radiator. If the box can be maintained below 60 W, this configuration can be cooled by conduction through board attachment devices called wedgelocks and down the chassis sidewalls down into the interface to the radiator. Power greater than 60 W usually requires chassis conduction enhancement, such as the use of thermal straps, pyrolytic graphite conduction bars, and heat pipes. Figure 1.4 shows the complicated electronics and wiring layout onto the heat spreader plate of the Curiosity rover. This particular heat spreader plate, called the rover avionics module plate (RAMP), is liquid cooled through serpentine fluid channels that take the heat out to heat radiators on the backside of the rover.

With the advent of CubeSats and miniaturized electronics [4] that has ushered in the New Space movement, the spacecraft level has been all but eliminated, or at least drastically reduced in size. The spacecraft level subsystems, including attitude control system (ACS), propulsion, power, and communications, are being integrated into Level 2 (see Figure 1.5), and thus a merging of the spacecraft and the EP engineer roles has taken place, or both roles have been eliminated and instead changed to the systems engineer role.

The New Space movement demands low-cost access to space, which includes eliminating redundant engineering roles and merging overlapping disciplines.

TINKERER VS. ENGINEER [5]

Because New Space demands quick access, sometimes companies will eliminate the engineering process altogether and go right to testing of the electronics based on a set

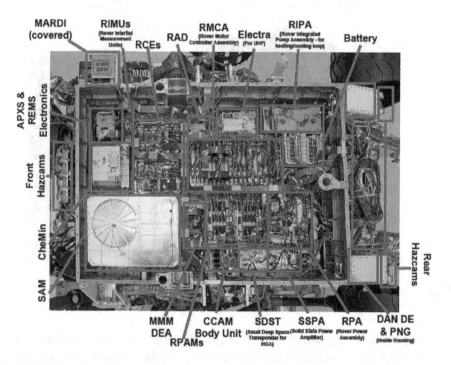

FIGURE 1.4 Packaging is important because of the limited volume in spacecraft. This top view of Curiosity rover shows all of the electronics and wire harnessing.

(Source: https://llis.nasa.gov/lesson/11201)

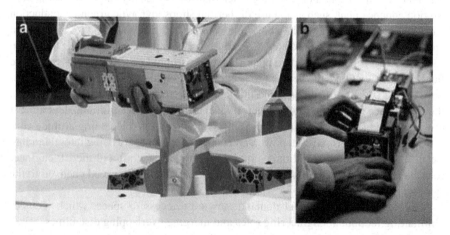

FIGURE 1.5 CubeSat spacecraft [4].

of standards. If they pass, then they are assembled, placed on a rocket, and launched, with fingers crossed. As much as standards help us selectively choose what stands a chance of surviving in space, they do not replace engineering first principles, which is what separates the engineer from the tinkerer. First principles involves looking at a problem from its fundamentals, which for engineering involves analyzing the product against the basic spacecraft killers—launch loads, thermal loads, changes from the environment, and repetitive cycles such as vibrations during launch. Engineers have tools that can aid in analysis, including computer-based modeling, though many of the basic analysis can be done by hand on a piece of paper and with a calculator or with nothing more than an Excel spreadsheet. The analysis not only helps catch potential weak points or mistakes in a design, but it can help in arriving at the best, most SWaP-effective solution, which in the end helps save time and money. The importance of knowing electronic packaging engineering principles cannot be overstated, as they guide the success of the program.

BAKED-IN RELIABILITY

Reliability needs to be designed into the product early on, or euphemistically "baked in," and cannot just simply be applied at the back end during vibration and/or thermal vacuum tests. Legacy companies have the benefit of drawing upon years of flight heritage, whether it is proven hardware or battle-tested key staff with many years of experience [6]. New Space companies, usually existing as startups, do not have the old sage with 30 years of experience who knows what works and what does not. These companies need to focus on designing reliability into the product, which often requires strict adherence to standards, many of which are found in the last chapter of this book and openly available to anyone who wants to place reliability electronic hardware in space.

REFERENCES

1. Subramanya, P., Jiwan Kumar Pandit, C.S. Prasad, and M.R. Thyagara, "Vibration analysis study of spacecraft electronic package: A review," *International Journal of Science, Engineering and Technology Research (IJSETR)*, vol. 3, no. 3, March 2014.
2. Prokop, Jon S., Peter Sandborn, and Kristan Evans, "The application of high density electronic packaging for spacecraft cost and mass reduction," https://doi.org/10.1016/0094-5765(94)00248-K
3. Schaefer, Edward D., Vincent L. Bailey, Carl J. Ercol, Sharon X. Ling, Ron C. Schulze, and Steven R. Vernon, *Spacecraft Packaging*. The Johns Hopkins University Applied Physics Laboratory MESSENGER website, http://messenger.jhuapl.edu [cited August 18, 2008].
4. Papadopoulos, Evangelos, Iosif Paraskevas, and Thaleia Flessa, *Miniaturization and Micro/Nanotechnology in Space Robotics*. NanoRobotics: Current Approaches and Techniques.
5. www.nsta.org/science-and-children/science-and-children-february-2020/tinkering-vs-engineering
6. www.intelligent-aerospace.com/military/article/14201701/air-force-researchers-able-to-reduce-the-size-weight-and-power-swap-for-rf-and-microwave-components

2 New Space

The launching of Sputnik by the Soviet Union ushered in the Space Race in 1957 at the height of the Cold War, and it was a matter of national security that prompted the United States to answer back with its first spacecraft in 1958, called Explorer 1, which was created by NASA's Jet Propulsion Laboratory (JPL) [1–15]. At JPL, a replica of the spacecraft sits in the vestibule of their museum and the Von Karman auditorium, which together house replicas of most of JPL's most famous spacecraft, including the Mars exploration rovers and the Cassini, Galileo, and Voyager spacecraft. The Galileo spacecraft in the museum is a one-to-one scale replica, which stands about five meters high from thrusters to high-gain antenna, and its massive magnetometer extends out 20 meters, taking up a good portion of the museum's ceiling space. The Von Karman auditorium next door houses a large Cassini spacecraft replica, but only at half-scale, and one is left with the wonder of how large the real thing must have been. This was the era of large school-bus-size spacecraft, which were measured in thousands of kilograms, a far cry from today's New Space spacecraft.

Miniaturization has made constellation projects more possible and is popularized by SpaceX's Starlink program, which has hundreds of small spacecraft orbiting at a relatively low earth orbit to give broadband coverage across most of the globe. New Space seeks to make space accessible to more people by bringing down the cost of sending hardware into space to do real-time global imaging, or real-time Google Maps such as Planet is doing, or synthetic aperture imaging such as Cappella to have high-resolution imaging not only in the day, but at night.

I once visited the company that made the models that went into the JPL museum, as well as in many space museums across the country. The name of the company is Scale Models in Hawthorne, CA, the same city in which SpaceX headquarters is located. The owner once told me that they purchase and use the same radiators and spacecraft insulation wrapping, known as multilayer insulation (MLI), that is used in the real spacecraft to make it look as authentic as possible. The company mostly made scaled-down models of the spacecraft, which leads me to think that these scaled-down versions, sometimes down to 1/10th scale, are closer to the newest types of spacecraft going into space today. The advent of the CubeSat, the 1U version of which is a 10 cm per side cube resembling a tissue box, has ushered in the latest space revolution called New Space. When I first started working at JPL in 2008, my first assignment was to help design the thermal subsystem for a black hole imaging spacecraft called NuStar. Although by JPL standards it was small at 350 kg, it had a massive 10-meter boom that separated the spacecraft bus with the optical bench assembly. The second spacecraft I helped design was a planet-finder spacecraft called Space Interferometry Mission (SIM), which was a rectangle with dimensions of 1 m × 1m × 6 m. The mission was disrupted by a much smaller-scale spacecraft called WFIRST, so SIM was ultimately canceled. This kind of disruption would start to become more commonplace, just as flat screen TVs disrupted

DOI: 10.1201/9781003247005-2

cathode-ray tube sets and are prevalent in today's small space industry. New Space not only reduces the cost of spacecraft by designing things smaller but also chooses to take risks by introducing commercial off-the-shelf (COTS) parts into the design, something the legacy aerospace companies are not allowed to do in part because of their own internal bureaucracy. Legacy companies include all the aerospace companies that were created at the beginning of the Space Race that led to the lunar landing of 1969 such as Northrup Grumman, Lockheed Martin, Boeing, NASA–Jet Propulsion Laboratory, Aerospace Corp., and many more. These companies traditionally are large and not too agile compared to SpaceX, for example, or the new kid on the block Rocket Lab, who have both disrupted the industry by drastically reducing the cost of access to space. Rocket Lab is uniquely positioned for success as they offer repeatable rocket launches currently about once a month, as well as satellite services including design, build, and operation. The New Space industry, including launch services, is set to see an enormous expansion from about $200 billion a year currently to $1.4 trillion by 2030 [16–18].

FOCUS ON RELIABILITY

The legacy companies all agree that one thing lacking from the new private space companies is reliability of their products. One researcher has estimated that 50% of the new space product sent up by these companies will fail prematurely or not work at all. With a push for faster, better, and cheaper, adherence to reliability principles takes a hit. This book ultimately looks to level the playing field by introducing legacy-style analysis and design of structural and thermal aspects of the spacecraft and applying them to the fast-paced small satellite market. Even during my time at JPL, many of us were pushing for lower-cost, high-performance technology to enable more projects and increase access to space. This book covers some of the latest technology that is scalable from a school-bus-size spacecraft to one that would fit in the palm of your hand.

FOCUS ON REPEATABILITY

New Space places a demand on repeatability of not only launch, but also what is launched. Constellation programs focus on putting hardware into space frequently and on demand as space companies look to replace the old with the new. The analysis and design principles demonstrated in this book will help the spacecraft engineer reduce the time of delivering flight hardware by predicting defects and designing them out of the system before they are built and tested, greatly reducing time in the thermal testing chambers or on the vibration tables.

FOCUS ON ROI (RETURN ON INVESTMENT)

By reducing hardware design and test turnaround times and identifying low-cost alternatives for thermal control and hardware ruggedization strategies, the cost of qualifying reliable and robust flight hardware goes way down. Getting hardware to the launch pad can be done quickly, without jeopardizing reliability and with less

FIGURE 2.1 Typical processor used for MSL M2020 etc. [18].

human effort, as running an analysis on a computer only takes one person versus a team working around the clock during a thermal vacuum testing campaign. Another way of improving ROI is to reuse assemblies across many spacecraft platforms, such as using the RAD 750 flight computer slice by BAE Systems (Figure 2.1), which are 3U and 6U in size. For smaller spacecraft such as the Mars Helicopter by JPL, a COTS assembly called the Snapdragon (Figure 2.2) offer performance at a smaller footprint. These assemblies are allowed on projects that are class D or E that are high risk, are higher reward, and have a limited life expectancy.

New Space demands innovative, out-of-the-box thinking and thinking on one's feet, made popular by innovative scientists and engineers at JPL but tailored for this new fast-paced and low-cost environment. The New Space revolution is pushing us towards reducing the turnaround time to put hardware into space. NASA and other legacy space organizations have cycle times from concept to finished product of several years, where new space is turning things around in a matter of months. Compromises often need to be made, and unfortunately trade-offs often focus on reducing lengthy testing campaigns that strive to demonstrate "test-as-you-fly" reliability principles. As will be shown in greater detail, reduced turnaround times do not have to reduce the reliability if the focus shifts away from lengthy test campaigns to adherence to design principles with emphasis on targeted analysis. A structural weak point caught during analysis and design instead of during the vibration test, for example, could save months of development and respin time. Likewise, with thermal,

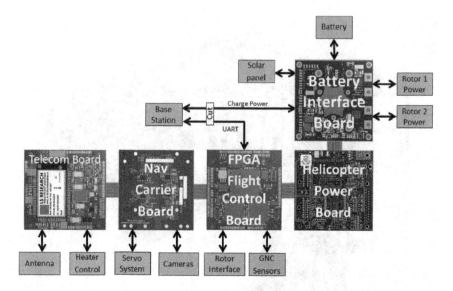

FIGURE 2.2 A good example of New Space actually comes from an old player. JPL used a COTS Snapdragon in the Mars Helicopter [18].

finding out that the hardware or spacecraft was not intelligently designed to remove the required heat could catastrophically set the program back several months to a year. Design and analysis are cheap compared to multiple thermal/vacuum test runs that can run several months at $5–$10 thousand per day.

REFERENCES

1. http://satellitemarkets.com/news-analysis/opportunities-emerging-new-space
2. "How hard is it to start a rocket company?" Ft. SpaceX & Rocket Lab, October 28, 2018, YouTube/curious_elephant
3. "SpaceX challengers—the great rocket race | MUST WATCH | Part 1," Start Up Division/YouTube, June 18, 2019.
4. "The new space race has officially begun—progress report," Freethink/YouTube, March 2020.
5. "Space: The final business frontier," Bloomberg Quicktake, October 22, 2019/YouTube (200B industry).
6. Bowler, Tim R., "A new space race," in *Deep Space Commodities*, Palgrave Macmillan, August 9, Singapor, Singapor, 2018, pp. 13–19.
7. "Launching from earth: The science behind space law and technological developments, Tom James Simon Roper," in *Deep Space Commodities*, Palgrave Macmillan, London, UK, pp. 21–51.
8. Shiroma, W.A., L.K. Martin, J.M. Akagi, et al., "CubeSats: A bright future for nanosatellites," *Central European Journal of Engineering,* vol. 1, 9–15, 2011, https://doi.org/10.2478/s13531-011-0007-8
9. Pomeroy, Caleb, Abigail Calzada-Diaz, and Damian Bielicki, *Fund Me to the Moon: Crowdfunding and the New Space Economy*, Space Policy Publisher: Elsevier Ltd., London, UK, February 2019.

10. Paikowsky, Deganit, "Original articles what is new space? The changing ecosystem of global space activity," *New Space*, vol. 5, no. 2, Published Online: June 1, 2017.

11. Denis, Gil, Didier Alary, Xavier Pasco, Nathalie Pisot, Delphine Texier, and Sandrine Toulza, "From new space to big space: How commercial space dream is becoming a reality," *Acta Astronautica*, vol. 166, 2020, pp. 431–443.

12. "What does it take to compete in NewSpace," 2016, www.forbes.com/sites/saadiampek-kanen/2016/06/28/what-does-it-take-to-compete-in-newspace/#7335281041f8.

13. "The NewSpace revolution: The emerging commercial space industry and new technologies," 2017, www.geospatialworld.net/article/emerging-commercial-space-in-dustry-new-technologies/

14. "More ISP competition is the key to restoring net neutrality," March 2018, www.entefy. com/blog/post/540/more-isp-competition-is-the-key-to-restoring-net-neutrality.

15. www.jpl.nasa.gov/who-we-are/history

16. "Satellite small market research," Market Research Engine, September 14, 2018, www. marketresearchengine.com

17. Rocket Lab Investor Information, March 2021, www.rocketalbusa.com

18. https://rotorcraft.arc.nasa.gov/Publications/files/Balaram_AIAA2018_0023.pdf; www. extr emetech.com/extreme/149713-curiosity-swaps-out-its-primary-computer-to-hope fully-restore-full-functionality

3 Thermal/Structural Challenges in Miniaturizing

New space is also pushing the miniaturization of space hardware to increase turnaround times and decrease overall mass. Currently, Rocket Lab can put 300 kg at low earth orbit for a launch price of $7 million, or $23.3 thousand per kilogram, or roughly half of the cost of the payload's weight in gold. Traditionally, a rule of thumb was that launch will cost the payload's weight in gold, and even though this price is now cut in half, there are dramatic savings to be made by reducing overall mass. This reduction in mass comes at a price. Thermal does not scale linearly with mass, or in this case let us assume volume. This is especially evident with the advent of CubeSats, where electronics are crammed into a 1U cube 10 cm in length. If x is the length of one side of the hardware cube, for example, power density will increase by x^3, which means the denser the electronics are packed, the ability to cool them will increase by the cube of the length. To remediate this thermal concern, an organization to promote a high-bandwidth form factor called SpaceVNX was created.

AN ASSESSMENT OF SPACEVNX BY JPL ELECTRONICS PACKAGING GROUP

The document VITA 74 Draft Standard [1] was reviewed by Jeremiah and Juan to see if there were any packaging concerns per JPL requirements. Supporting literature reviews included a statement from https://www.trident-sff.com/ about what is now SpaceVNX [2–5]. A report created by a JPL transceiver was translated into 3U VPX format that is cross compatible with VNX. It was mentioned that the transceiver dissipated 20 W, and like VNX format had both base and mezzanine cards and the Xilinx FPGA on the mezzanine that required a heat pipe for cooling. The VPX design calls for card wedgelocks for thermal performance and structural support. It was believed that JPL would be much better served designing a VPX chassis with VPX modules before attempting a VNX system.

The VNX design (Figure 3.1) does not specify wedgelocks and instead identifies two heat spreaders with at least three thermal interface material callouts per module. By looking at the VNX draft specification, it was unclear how the heat would be removed from the module via the heat spreaders without the use of wedgelocks, though a discussion with a VPX consortium member, who also sits on the SpaceVNX consortium, mentioned the existence of updated models that demonstrate how heat is conducted from the printed wiring assemblies (PWAs) to the outer chassis (Figure 3.2). Juan attempted to do a simple thermal analysis that showed that heat

DOI: 10.1201/9781003247005-3

FIGURE 3.1 Example of the VNX 1U form factor.

Mounting – Solid Plate
interface to S/C radiator
panel or RAMP

Sidewalls –
Solid walls. May
require APG material,
or heat pipe
enhancement

FIGURE 3.2 Conversion to SpaceVNX (conduction cooled) VITA 74.4.

would need to be removed along the sides where the wedgelocks normally reside and the "handle side" of the module, which is the face opposite the backplane connectors (Figures 3.3–3.5). In our discussion, this was the understanding that the newer VNX models demonstrated. It was apparent that heat conduction between modules was not acceptable.

FIGURE 3.3 1.2–3 VITA 74 module without cover showing axis of relative orientation to cube.

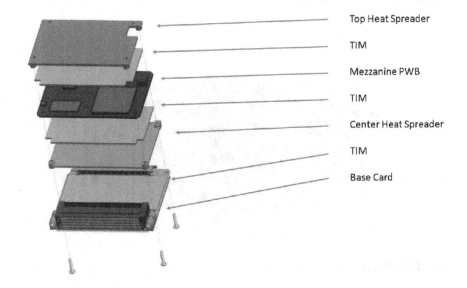

Top Heat Spreader

TIM

Mezzanine PWB

TIM

Center Heat Spreader

TIM

Base Card

FIGURE 3.4 1.2–4 VITA 74 module exploded view.

Also unclear was the structural rigidity of the SpaceVNX design since it was missing wedgelocks to bind the assembly to the chassis. There was not enough information in the draft specification to show that the modules would meet JPL requirements for a given launch vehicle.

Going forward, it was mentioned that it was possible to deliver the new model to JPL to test against thermal and structural requirements. Also available for purchase was an existing printed model that could be sent to JPL for demonstration.

The current chassis design is finned to aid in cooling, but for space applications it will have to be changed to a conduction cooling approach—one option is shown in Figure 3.2. Heat pipes can be embedded in the sidewalls to promote a balanced heat

FIGURE 3.5 Proposed small module size.

FIGURE 3.6 Proposed larger module size.

spreading without incurring a large gradient from one side of the cube to the other in the z-direction.

Each slice is depicted in Figure 3.3, and a daughter card and a mezzanine board sandwiched between two heat spreads are depicted in Figure 3.4. The approach has three thermal interface materials (TIMs) and two heat spreaders. Heat conduction from boards to chassis must primarily happen through the spreaders in the y-direction.

The SpaceVNX module comes in two sizes, as shown in Figures 3.5 and 3.6. The module and chassis assembly are shown in Figure 3.7. What is unclear is how to transfer heat from the heat spreaders to the walls of the modules and out to the chassis walls. The two-size modules would need to have a shell that could integrate well with the internal heat spreaders, or be the heat spreaders themselves, to reduce the temperature gradient between module boards and chassis wall. Also, there are

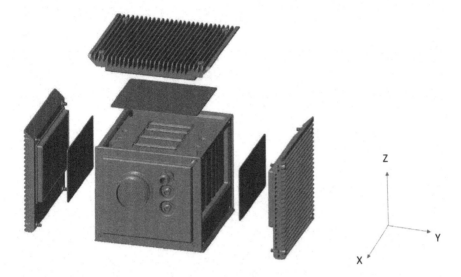

FIGURE 3.7 VITA 74 Chassis design, exploded view.

too many TIMs in the thermal path between devices and chassis walls. Those TIMs will need to be of high performance to manage thermal gradients.

REFERENCES

1. VITA 74 Draft Standard 1 VITA 74.0 2 VITA 74 Compliant System Small Form Factor Module 3 Base Standard 4, July 6, 2017.
2. What is SpaceVNX? www.trident-sff.com/
3. A SpaceVPX-compliant transceiver for low-SWaP instrument applications PPT presentation.
4. www.synqor.com/vpx/vpx_caseoutlines.html
5. Relevant Websites: www.cisco.com/c/en/us/td/docs/solutions/GGSG-Engineering/Dusty/ hardware/guide/Cisco_5940_Embedded_Services_Router_Hardware_Technical_Refer ence_Guide.html; http://advancetech.in/rugged-power-supplies/dc-dc-converters-3u-cpci

4 Fundamentals of Heat Transfer by Conduction and Convection

INTRODUCTION

To best understand how electronic packaging analysis is performed, the principles need to be identified and understood, especially the fundamentals of heat conduction and convection for estimating temperatures. Once in space, electronics primarily rely on conducting heat out and into cold space, since convection requires a cooling medium such as air or water. However, we can design systems in which we take the cooling medium with us, such as pumped fluid loops, so convection then becomes a prime mode of heat transfer. The third mode of heat transfer is radiation, which requires its own chapter to fully explain. This chapter is boiled down to the fundamentals, and the reader is encouraged to consult the referenced textbooks for a better understanding.

CONDUCTION HEAT TRANSFER

Conduction heat transfer is characterized by **Fourier's Law** of heat conduction for one-dimensional heat conduction [1] as:

$$\dot{Q} = -kA\frac{dT}{dx} \tag{4.1}$$

where k is the thermal conductivity of the material, dT/dx is the temperature gradient, and A is the cross-sectional area.

Heat is conducted in the direction of decreasing temperature, and thus the temperature gradient is negative when heat is conducted in the positive x-direction.

As an approximation and assuming thermal conductivity is constant for a given temperature range ΔT and A is constant, Fourier's law can be approximated as:

$$\dot{Q} = -kA\frac{\Delta T}{\Delta x} \tag{4.2}$$

Where ΔT = temperature difference across a distance Δx. In space where there is no atmosphere to help cool electronic devices, the electronics primarily need to be conduction cooled, with radiation as the secondary cooling mechanism.

DOI: 10.1201/9781003247005-4

CONVECTIVE HEAT TRANSFER

The bulk of convective cooling in space is the used of pumped fluid loops to move heat around the spacecraft, both for cooling hot components and for heating for temperature control during eclipses in orbit or for nighttime temperature buffering at night on Mars. On occasion, the thermal analysis may be asked to predict temperatures on the launch pad aided by the availability of air, and that would involve natural convective currents of air to cool components.

The convective equation is deceivingly simple [2]:

$$\dot{Q} = hA\left(T_h - Tc\right) \tag{4.3}$$

Where $\dot{Q}$ is the heat transfer rate in watts, h is the heat transfer coefficient in $\frac{W}{m^2}$, T_h and Tc are hot and cold temperature, respectively, though one of these temperatures represents the bulk fluid and one represents the device. The difficult part in this calculation is to derive a good *heat transfer coefficient*, h. A very rough but easy-to-remember rule of thumb for h ranges are as follows [2]:

In h units of $\frac{W}{m^2}$,

Stagnant gas convection, (e.g., stagnant air), $1 < h < 10$
Force gas convection $10 < h < 100$
Stagnant liquid convection (immersion) $100 < h < 1000$
Forced liquid convection $1000 < h < 10,000$
Boiling and phase change $10,000 < h < 100,000$

Again, there is some overlap as a fluid with non-optimal thermal properties and/or with low flow, which can behave nearly like a stagnant fluid, but the ranges are meant to give the reader a range of what is possible.

To arrive at a better estimate of the convection heat transfer coefficient, correlation of the Nusselt number, Nu, are usually explored and has the relationship to h as follows:

$$Nu = \frac{hL_c}{k} \tag{4.4}$$

Where the Nusselt number is dimensionless and gives an estimate of the effectiveness of the fluid to remove heat by convection versus conducting heat away:

$$\frac{\dot{Q}_{conv}}{\dot{Q}_{cond}} = \frac{h\Delta T}{k\Delta T/L} = Nu \tag{4.5}$$

The reader is encouraged to look at the various estimations of Nu in the convection heat transfer in this section, though calculation of Nu is usually accompanied by two more dimensionless numbers, the Reynolds number, Re, and the Prandtl number, Pr,

$$Re = \frac{\rho V L_C}{\mu} \tag{4.6}$$

Where L_c is the characteristic length (plate surface length, diameter, etc.), r is the density, V is the velocity, and m is the dynamic viscosity. Reynolds number shows the ratio of the fluid's inertial force to viscous force, and:

$$Pr = \frac{\mu c_p}{k} \tag{4.7}$$

indicates the fluid's molecular diffusivity of momentum/molecular diffusivity of heat. Here k is the conductivity of the fluid, and c_p is the specific heat.

Again, Nu, Pr, and Re are usually depicted in the following format [3]:

$$Nu = C Re_L^m Pr^n \tag{4.8}$$

Where m and n are constant exponents commonly between 0 and 1 and C is a constant that depends upon geometry. The following are general Nusselt number approximations. For more precise estimates, refer to the textbooks referenced at the end of the chapter.

For turbulent flow in s simple tube, we have:

$$Nu = 0.0243 Pr^{0.4} Re^{0.8} \tag{4.9}$$

For laminar flow, the Nusselt number simplifies to:

$$Nu = 3.66$$

For turbulent flow over a plate:

$$Nu = 0.029 Pr^{0.43} Re^{0.8} \tag{4.10}$$

For laminar flow over a plate:

$$Nu = 0.332 Rex^{1/2} Pr^{1/3}, \text{for } Pr > 0.5 \tag{4.11}$$

REFERENCES

1. Cengel, Yunus A., *Heat and Mass Transfer*, 5th Ed., McGraw-Hill, New York, 2007, pp. 18–26.
2. Gieck, R., and K. Gieck, *Engineering Formulas*, 6th Ed., McGraw-Hill, New York, 1990.
3. Mills, A.F., *Heat Transfer*, 2nd Ed., Prentice Hall, Upper Saddle River, NJ, 1999, pp. 301–315.

5 Fundamentals of Heat Transfer by Radiation

INTRODUCTION

The only true way to cool electronics in space ultimately is by radiation. Though electronics are often analyzed by conservatively ignoring radiation at the electronics level, as spacecraft miniaturizes, the radiation panels likely will transition from the spacecraft level to the chassis level. The packaging engineer will need to be well informed about radiation heat transfer and how to best manipulate surface properties, tapes, and paint coatings.

All matter continuously emits electromagnetic radiation, which travels $c_0 = 3 \times 10^8 \ m/s$ through a vacuum at the speed of light [1]. Radiation exhibits both wave nature and particle nature. The wavelength of radiation λ is related to its frequency v_f and speed of propagation c as:

$$c = v_f \lambda \tag{5.1}$$

The units used for λ are micrometers, microns (1 μ = 10⁻⁶ m), meters, or angstroms (10⁻¹⁰ m). The unit used for hertz (1 Hz = 1 s⁻¹). The wavenumber v of the radiation is related to its wavelength and frequency as:

$$v = \frac{1}{\lambda} = \frac{v_f}{c} \tag{5.2}$$

The units for v are usually cm⁻¹; hence $v \left[cm^{-1} \right] = 10^4 / \lambda \left[\mu m \right]$. Figure 5.1 shows the *electromagnetic spectrum*. Thermal effects are associated with radiation in the band of wavelengths from about 0.1 to 100 μm, and visible radiation is in the very narrow band from about 0.4 to 0.7 μm.

According to quantum mechanics, radiation interacts with matter in discrete quanta called photons, with each photon having an energy E given by:

$$E = h v_f \tag{5.3}$$

Where $h = 6.626 \times 10^{-34} \ J \cdot s$ is *Plank's constant*. Each photon also has a momentum p given by:

$$p = \frac{h v_f}{c} \tag{5.4}$$

DOI: 10.1201/9781003247005-5

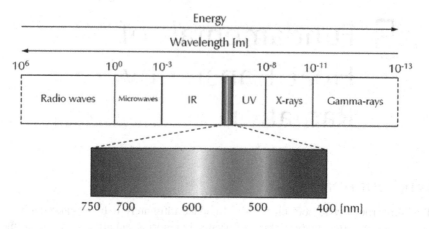

FIGURE 5.1 Electromagnetic spectrum.

In most solids and liquids, the radiation emitted by one molecule is strongly absorbed by surrounding molecules. Thus, the radiation emitted or absorbed by these liquids or solids involves only a layer of molecules close to the surface; for metals, this layer is a few molecules thick, whereas for nonmetals, it is about a few micrometers thick. For these materials, radiation emission and absorption can be regarded as *surface* phenomena. On the other hand, in gas mixtures containing species such as water vapor and carbon dioxide, or in a semitransparent solid, the absorption is weak. Radiation leaving the body can originate from anywhere in the body. Emission and absorption of radiation are then *volumetric* phenomena. Radiation transfer between surfaces through a nonparticipating medium is relatively simple to analyze.

BLACKBODY RADIATION

A blackbody emits energy over a broad spectrum of wavelengths. The energy is emitted in all directions from a surface. The intensity ($I_{\lambda,b}$) or "brightness" of the emitted radiation is a function of wavelength λ, but is independent of direction θ and φ. The emissive power of a blackbody obeys Lambert's cosine law, relating it to intensity [2]. T[3]he emissive power is a function of direction (Figure 5.2):

$$I_{\lambda b} = \frac{2hc_0^2}{\lambda^5 \left[\exp\left(hc_0/\lambda kT\right)-1\right]} \tag{5.5}$$

Integration of radiation intensity over a hemisphere surrounding a blackbody surface for the spectral emissive power of a blackbody is shown in Figure 5.3:

$$E_{\lambda b} = \int_0^{2\pi} \int_0^{\pi/2} I_{\lambda b} \cos\theta \sin\theta \, d\theta \, d\phi \tag{5.6}$$

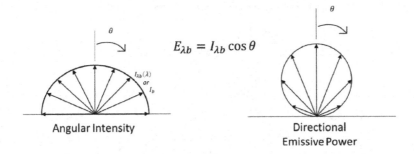

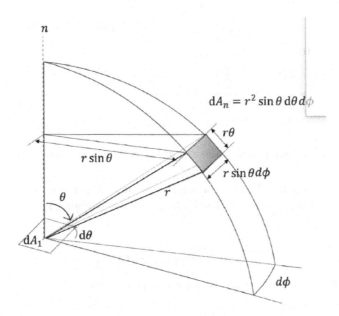

FIGURE 5.2 The intensity of radiation ($I_{\lambda,b}$) as a function of wavelength is predicted by Planck's law.

FIGURE 5.3 Integration of radiation intensity over surrounding hemisphere.

Cos θ term is from the projected differential area on a hemisphere and sin θ term is from a differential solid angle:

$$E_b = \pi \int_0^\infty I_{\lambda b}\, d\lambda = \sigma T^4 \tag{5.7}$$

$$E_{\lambda b} = \pi I_{\lambda b} \tag{5.8}$$

Finally integrating the spectral emissive power over all wavelengths gives the total hemispherical emissive power of a blackbody (aka the Stefan-Boltzmann law).

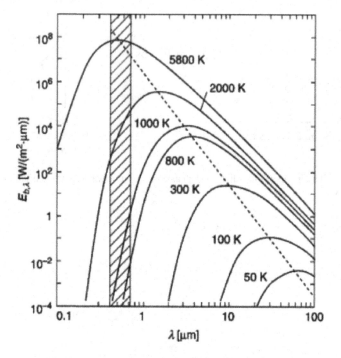

FIGURE 5.4 Blackbody spectral density showing emissive power vs. wavelength.

Emissive power is a function of wavelength and temperature for a blackbody. Emissive power is predicted by Planck's law, as shown in Figure 5.4:

$$E_{\lambda b} = \frac{2h\pi c_0^2}{\lambda^5 \left[\exp\left(h c_0 / \lambda kT \right) - 1 \right]} \tag{5.9}$$

Wein's law predicts spectral peak for given temperature.

EMISSIVITY

Emissivity compares radiant energy emitted from a real surface to that of a blackbody at the same temperature [3]. Real surfaces emit less energy at a given temperature, wavelength, and direction than a blackbody does. Emissivity depends only on the physical properties of a surface and its temperature; it is usually measured normal to a surface and as a function of wavelength (Figures 5.5–5.7). Computing energy loss from a surface requires integrating emissivity over all directions, and most metals have low emissivity unless they are oxidized. Most nonmetals have high emissivity, approaching that of a blackbody [4]. Emissivity can have variations with direction, wavelength, and temperature, as shown in Figure 5.5, and representative directional distributions of total directional emissivity for conductors and nonconductors. Conductors tend to increase emissivity with angle while nonconductors tend to decrease emissivity with angle.

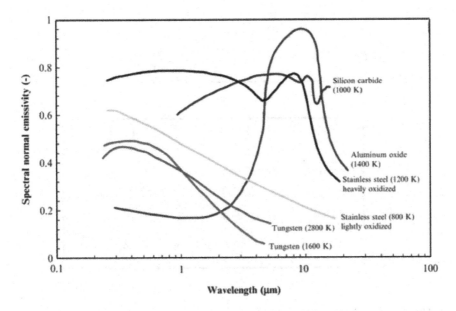

FIGURE 5.5 Spectral dependence of emissivity of select materials.

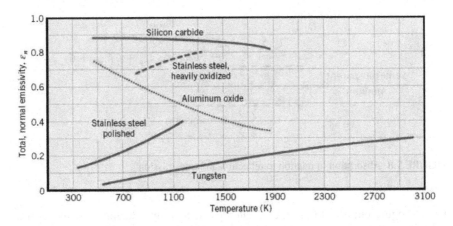

FIGURE 5.6 Temperature dependence of emissivity for select materials. Some materials have significant changes with temperature, so be careful when using tabulated data.

ABSORPTIVITY

Absorptivity as used in spacecraft design is the ability of the surface to absorb solar radiation (Figure 5.8). In most cases, absorptivity wants to be kept to a minimum; however, some designs may want to take advantage of solar heating to perform a function [5]. Also, some high-absorptive surfaces are needed; for example, black paint to reduce glint around the camera field of view, so the packaging engineer will

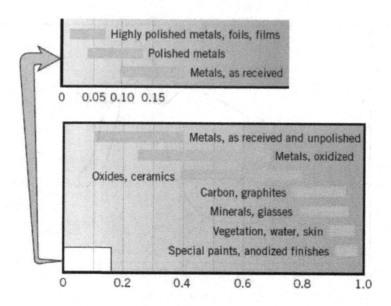

FIGURE 5.7 Total, normal emissivity.

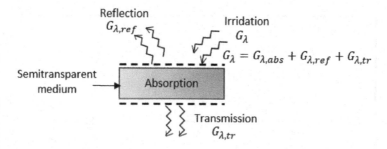

FIGURE 5.8 Absorption in relation to reflection and transmission.

need to figure out how to use these materials without introducing excessive heat to the electronics.

Engineers typically work with surfaces with selective spectral absorption. Directional dependence is averaged out to hemispherical absorption, and absorptivity is the fraction of incident radiation absorbed by a surface. It is a function of source temperature, incident direction, wavelength, and the physical properties of all surfaces involved in radiant heat exchange (Figure 5.9).

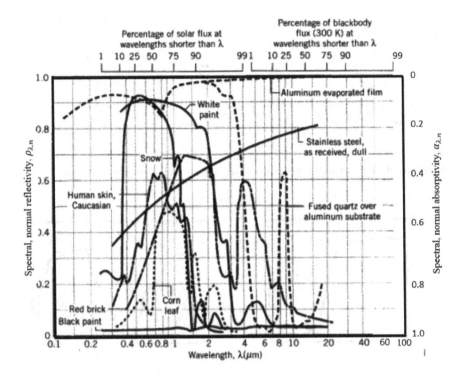

FIGURE 5.9 Spectral dependence of spectral, normal absorptivity and reflectivity of selected opaque materials.

REFLECTIVITY

For opaque surfaces, what is not absorbed is reflected. Reflections can be specular, diffuse, or in between; the size of arrows in Figure 5.10 indicates the directional emissive power of reflection.

GRAY SURFACES

Real surfaces absorb and emit less radiation than black surfaces do; the fraction of energy that is absorbed is called the absorptivity and is given the symbol α. If the solid is opaque, the fraction of energy reflected is called the reflectivity, ρ [6]. Note that $\rho + \alpha = 1$. If the solid is transparent, the portion of energy that passes through the solid is given by the transmissivity, τ. Note that $\rho + \alpha + \tau = 1$. The emissivity is the fraction of thermal radiation emitted compared to blackbody emission, and for gray surface assumption the radiative properties of absorptivity and emissivity do not vary with wavelength over the spectral region of the radiation sources and the surface emission. For a gray surface, the absorptivity and emissivity are equal, $\varepsilon = \alpha$ (Figure 5.11).

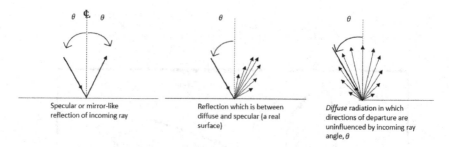

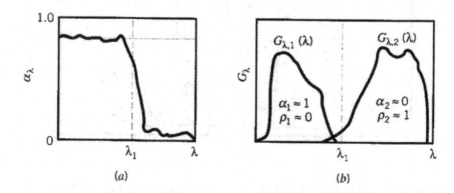

FIGURE 5.10 Directional emissive power of reflection based on surface properties.

FIGURE 5.11 Spectral distribution of (a) the spectral absorptivity of a surface and (b) the spectral irradiation at the surface.

The left figure is the surface spectral absorptivity distribution and the right figure shows two different spectral irradiation distributions on the surface. The surface is not gray because α is different for the two different irradiation sources. Emissivity of the surface remains the same for the surface. Here $\alpha \neq \varepsilon$ over the spectral regions (Figure 5.12).

This is a gray surface. Irradiation and surface emission are in a spectral region where spectral properties are approximately constant (in top figure). The middle figure shows irradiation field of source. The lower figure shows surface emissive power.

DIFFUSE SURFACES

The intensity of emitted radiation is independent of direction for diffuse surfaces. A blackbody is an ideal diffuse emitter. A blackbody is the ideal standard for directional properties of surfaces, while a cosine-law surface is [7–10] a diffuse surface. Diffuse means that directional emissivity and absorptivity are independent of direction, while diffuse-gray surfaces are simplifications to remove directional and wavelength dependence on surface properties.

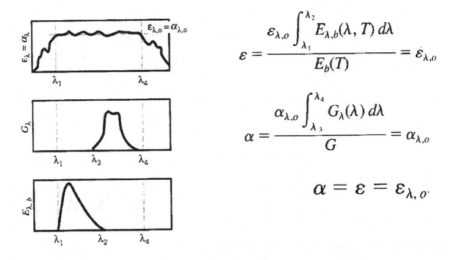

$$\varepsilon = \frac{\varepsilon_{\lambda,o} \int_{\lambda_1}^{\lambda_2} E_{\lambda,b}(\lambda, T)\, d\lambda}{E_b(T)} = \varepsilon_{\lambda,o}$$

$$\alpha = \frac{\alpha_{\lambda,o} \int_{\lambda_3}^{\lambda_4} G_\lambda(\lambda)\, d\lambda}{G} = \alpha_{\lambda,o}$$

$$\alpha = \varepsilon = \varepsilon_{\lambda,o'}$$

FIGURE 5.12 A set of conditions for which gray surfaces behavior may be assumed.

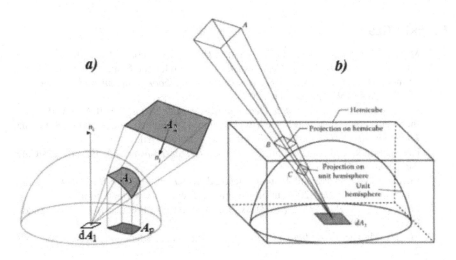

FIGURE 5.13 (a) The unit-sphere method and (b) its comparison with the hemicube method (Howell et al.).

RADIATION GEOMETRY

Radiation emitted by a surface propagates in all directions, and sometimes we need to know its directional distribution dA_I is a differential area emitting radiation. dA_n is a differential area element on the hemisphere through which the radiation passes $dA_n = r^2 sin\theta d\theta d\varphi$. The solid angle through dA_n is defined by $d\omega = dA_n/r^2$ or $d\omega = sin\theta d\theta d\varphi$ (see Figures 5.13–5.14) [11–12].

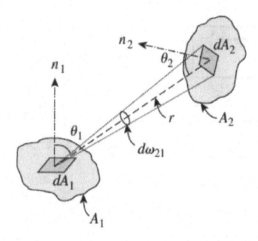

FIGURE 5.14 Radiation transfer from dA_1 to dA_2.

For the case where the dA_n represents the area normal to the radiation, we have $dA_n = dA_1 \cos \theta$.

REFERENCES

1. Mills, A.F., *Heat Transfer*, 2nd Ed., Prentice Hall Inc., Concord, MA, 1999.
2. Edwards, Donald, *Radiation Heat Transfer Notes*, Taylor and Francis, New York, 1981.
3. Gilmore, David G., *Spacecraft Thermal Control Handbook: Fundamental Technologies*, 2nd Ed., AIAA, Washington, DC, 2007.
4. Holman, Jack P., *Introduction to Heat Transfer*, 9th Ed., McGraw-Hill, New York, 2001.
5. Incropera, Frank P., David P. DeWitt, Theodore L. Bergman, and Adrienne S. Lavine, *Fundamentals of Heat and Mass Transfer*, 6th Ed., John Wiley, New York, 2007.
6. Kreith, Frank, and Mark Bohn, *Principles of Heat Transfer*, 6th Ed., Brooks/Cole, Pacific Grove, CA, 2000.
7. Lienhard IV, John H., and John H. Lienhard V, *A Heat Transfer Handbook*, 3rd Ed., Cambridge, MA, 2000–2006, http://web.mit.edu/lienhard/www/ahtt.html.
8. Mills, Anthony, *Basic Heat and Mass Transfer*, 2nd Ed., Prentice-Hall, Upper Saddle River, NJ, 1998.
9. Modest, Michael F., *Radiative Heat Transfer*, 2nd Ed., Academic Press, New York, NY, 2003.
10. Siegel, Robert, and John R. Howell, *Thermal Radiation Heat Transfer*, 4th Ed., Taylor Francis, New York, NY, 2001.
11. Sparrow, E.M., and R.D. Cess, *Radiation Heat Transfer*, Augmented Ed., HarperCollins, New York, NY, 1978.
12. Cengel, Y.A., *Heat and Mass Transfer: A Practical Approach*, McGraw-Hill, New York, NY, 2018.

6 The Multilayer Insulation (MLI) Blanket

Capstone spacecraft for moon mission (Rocket Lab).

BLANKET APPLICATIONS

MLI blankets are commonly used for all aspects of the spacecraft, from the highest level of thermal control to the hardware level, especially for cryogenic, Radio Frequency (RF), and optical systems where tight thermal control is required, as well as for where small heat leaks and parasitic loads can be tolerated [1–7]. Non-flight engineering models (EM) have different restrictions as they will not see flight and help to control the thermal environment during a test to maintain tight temperature control. A common mistake is to assume that methods allowed on an EM blanket, such as taping instead of stitching, is acceptable for flight. Relaxing the rules for EM blankets helps speed up the test, avoiding paperwork associated with flight hardware, as well as making things such as grounding easier to accomplish. The understanding, however, is that the flight-set will still have to meet the necessary restrictions, and thus time and money should be allocated appropriately.

DOI: 10.1201/9781003247005-6

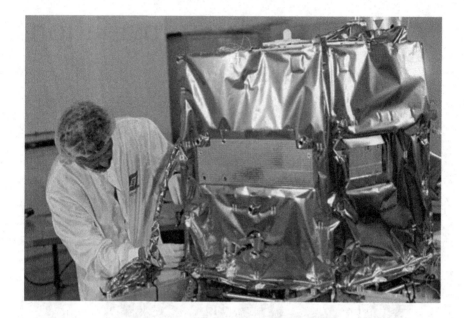

FIGURE 6.1 Typical spacecraft multilayer insulation (shown in gold).

(Source: https://blenderartists.org/t/satellite-mli-blankets-modeling/1179559)

Assembling mechanical ground support equipment can be a large effort with high cost, so alleviating restrictions, such as simplifying attachments, can reduce cost and setup time.

DESIGN (DRIVEN BY REQUIREMENTS AND CONSIDERATIONS)

The purpose of the MLI blanket is to minimize heat loss from the hardware to space. Since power is typically a scarce resource, heater sizes and power requirements can be reduced to a minimum with a good blanket design. The blanket will also help reduce spacecraft sensitivity to changes in the space environment, including solar flux, albedo, Infrared (IR) flux from earth or other planets and moons, rocket plumes, and aero heating. The protective layers also help to protect against micrometeoroid damage and are often part of the mitigation plan. Standoffs are created by 5–7 mil aluminized Mylar around propellant tanks, which are the most sensitive to micrometeoroid damage. Also, effective protection is the use of one to two layers of beta cloth, though its greater density cautions the user to apply it only where necessary. The blanket thermal performance depends on factors that include:

- Blanket size (seams to total area)—the smaller, the better
  - Effective emittance ~ 0.015 (10 layers, large blanket)
  - Effective emittance ~ 0.008 (20 layers, large blanket)

- Blanket compression—the looser, the better
- Blanket penetrations (cables, etc.)—the fewer, the better

Blanket layer count rarely exceeds 25 layers, and one achieves a point of diminishing return after about 20 layers, where the performance advantages flatten out and the extra layers are more difficult to justify due to the mass. Minimizing seams is also desirable as they introduce conduction, which is proportional with temperature difference with space.

MLI internal layer design. Alternating aluminum Mylar/Dacron net layers is standard lay-up. The Dacron, sometimes called scrim, prevents the Mylar layers from contacting each other and reduces heat loss through conduction. This blanket design yields the best-performing blanket (lowest effective emissivity, or e-star), but it is limited by Mylar's (at 250 °C) and Dacron's (at 256 °C) melting temperatures. For higher temperature resistance, embossed aluminized Kapton (at 430 °C) is a better option. The embossed Kapton MLI is also much easier to work with since the Dacron sheet is not needed, therefore it makes for a lighter overall blanket. However, the performance of the embossed blanket material is not as good as Dacron/Mylar, on average the effective emissivity is 20% higher for embossed MLI (Figure 6.3).

FIGURE 6.2 NASA's New Horizons spacecraft fully covered with a gold-colored MLI to retain internal heat.

(Source: https://en.wikipedia.org/wiki/New_Horizons)

FIGURE 6.3 Embossed MLI.

Hybrid designs. For resisting high temperatures due to things like engine plumes, high planet IR, or sun maneuvers, a hybrid lay up may be advantageous. A 20-layer blanket designed for heat resistance could be constructed of five layers of embossed Kapton plus 15 layers of aluminized Mylar.

MLI exterior layer. Optical properties often drive selection for the exterior layer, including resistance to handling, micrometeoroid impact, and stray light (glint). Exterior layer temperatures are managed with reflective material such as aluminized second-surface mirror (SSM) or SSM Kapton gold. The metalized coatings also offer reduced solar dependence. Drawbacks of using metallized outer layers include contamination and electrostatic discharge (ESD). Carbon-filled (black) Kapton is usually used to reduce glint, especially around cameras and other optics. Beta cloth (Figure 6.4), especially as a single layer, can offer the optical properties of white paint (high IR emissivity, low absorptivity) but is popular with International Space Station (ISS) missions where during-flight handling by placement equipment or by astronauts is an issue. Kapton as the outer layer is preferable over Mylar, as it is tear resistance and will hold up to handling.

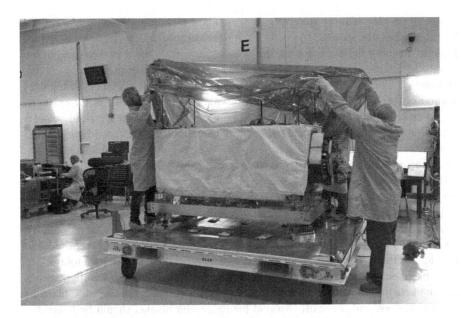

FIGURE 6.4 ECOSTRESS module that is attached outside the ISS shown with outer MLI surface covered in beta cloth.

(Source: jpl.nasa.gov)

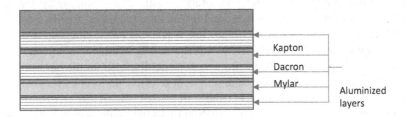

FIGURE 6.5 Typical MLI layers and patterning.

MLI blanket edges. Blanket edges can be a problem, especially during handling, as the Dacron inner layer easily frays and Mylar layers can tear (see Figure 6.5). It is best to treat the MLI blanket as would a tailor assembling the fabric and lining of a suit. Edges should be sewn with Dacron thread for normal applications and Nomex or ceramic fiber thread for high-temperature applications. It is common to recruit and train MLI technicians from the garment and furniture upholstery industries. To ease the handing of MLI edges:

- Use hot glue to seal edges of MLI material temporarily.
- Bind edge with tape, and then sew the edges.
- Avoid the use of Velcro, though many organizations use it.

Recap on MLI blanket materials. In a standard alternating layer of aluminized Mylar and Dacron spacers, the outer layers are determined by need. First-surface Kapton, meaning the Kapton is on the outside, is typical; however, one can put the metalized coating on the outside if there are EMI concerns. For the outer layer, second-surface aluminized Kapton is common. Indium tin oxide (ITO) deposited onto Kapton is also used for cases with where atomic oxygen is a concern; for example, for orbits below 400 km. Carbon-filled Kapton on the outer layer, black in color, is used near or around cameras and optics to reduce glint. First-surface Mylar in combination with a Dacron net spacer, or separator, is most common. For high emissivity, silver-coated Teflon of at least 10 mils may be used, but it is most used on the radiator surface due to the low solar absorptivity. Beta cloth on the outer surface helps with low-reflective emissivity and toughness, and it has fire-retardant properties, especially near rocket plumes during launch.

Attachment of the MLI with lacing cords is recommended over tape or Velcro for reliability and cleanliness, and commonly Nomex lacing cords are preferred. For venting, the inner Mylar layers are offered perforated, which offers a path for trapped air to bleed out of the blankets during launch vehicle ascent. NASA usually requires the blankets to handle depressurization rates of 65 Torr/sec to assure the blankets will not balloon and come off. Around hardware, the blanket should be loose, but secure, to ease in venting. Interior/internal MLI layers are perforated (Aluminum (Al)-Mylar, Al-Kapton) while exterior MLI layers are not typically perforated.

The metallization in the blankets is usually beneficial for EMI and to help meet radiation targets, though ESD charge build-up can be a concern, so grounding of the blanket is crucial. Plasmas in mission environments tend to cause charge build-up during orbits around Earth, Jupiter, and Saturn as well. All layers of the MLI must be grounded to mitigate against charge build-up, and the ground is usually a common ground shared by hardware and payload. Teflon surfaces on the exterior are a nightmare to ground due to low conductivity and require special attention. A good blanket grounding will also prevent damaging electrostatic discharge to the hardware, and special discharge protection may be required for sensitive hardware. ITO coating on Teflon is helpful for achieving ESD goals.

Attachment methods. It is best to use Nomex lacing cords to attach the blanket to the hardware and spacecraft structure. Typically, two equal spaced holes that are 2 mm in diameter, spaced 6 mm apart, will help with looping the lacing cord. Planning the attachment methods ahead of time will reduce any rework to hardware. A less reliable method may be to epoxy down tie-down loops or clips that the lacing cord can be funneled through, especially where drilling holes is not feasible. Kapton or reflective thermal tape is acceptable for holding down the blanket to avoid surfaces such as radiators.

Creating patterns. It is common for blanket engineers to use CAD models to create an estimated pattern to fit spacecraft and hardware. A less effective way, especially for complicated geometries, is to take dimensions from a drawing. What may be more feasible is to create a 3D printed stereolithography model and use tape and paper to drape over the model and produce a flat pattern. Recesses and cut angles can be best estimated by patterning right off the hardware, or from the 3D model if the hardware does not exist. Patterning on 3D models can help identify cable and hose egresses and feedthroughs, as well as fit checks with adjacent hardware without interfering with spacecraft integration and test operations. Pattern development enables the following:

- Defining the attachment method
- Defining the ground points
- Identifying the seam lines
- Defining interfaces

Fabrication. The blanket materials are delicate, can build up charge when handling, and are prone to collecting dust particles and other contaminants. Proper ESD grounded tables covered in anti-static AMERSTAT (cleanroom curtain material) is preferred. Fabrication should be done in a cleanroom to prevent foreign object debris (FOD). Taping and sewing of the edges is preferred. Think of these blankets as a fine-tailored suit to best protect your expensive hardware or spacecraft.

Attachment features. Lace interface: This is the most durable and secure, but it is time-consuming and not efficient for small blankets. Taped interface: Use when there is no access for lace, and it is not a positive method of securing. Velcro: This provides for fast, repeatable installation, but it is not a positive method of securing and is a contamination source.

Bakeout. Bakeout reduces the risk of contamination prior to testing in a vacuum as well as lowers the risk of hardware contamination. Blanket materials tend to harbor water vapor, especially from the Dacron scrim. Typical bakeout duration may vary according to temperature, and the material provider should be consulted.

Installation and removal of moving interfaces. MLI must be well secured to prevent interference of moving elements such as motors and other mechanisms, which may include possible expansion of MLI as it transitions to a vacuum environment.

Handling. Exterior layers require durability due to repeated handling and removing during testing. The surface should be cleaned with ethyl alcohol since it is less intense than isopropyl alcohol. Reinforcement fibers are sometimes necessary in the outer layers to prevent tearing. Use of gloves should be used when handling blankets as fingerprints introduce contamination and alter the optical properties. To limit water vapor contamination, blankets should be maintained in an environment with less than 50% humidity.

Blanket cost. Factors that attribute to the cost of the blanket include:

- Regular time or overtime pay, second shifts, etc.
- Rushing and expediting orders
- Blanket patterns time and rework
- Blanket design and complexity
- Testing support, bakeout planning

Areas to avoid in the blanket design:

- Silicone tape adhesive can cause concerns, especially with optics
- Silver Teflon exterior layers when possible due to difficult grounding issues

- Velcro
- Damaging flight hardware during critical assembly and test
- Reduce sunlit blanket seams when possible
- Concave blanket surfaces that get sunlit due to concentrated solar energy

REFERENCES

1. Ross, Jr., R.G., *Quantifying MLI Thermal Conduction in Cryogenic Applications from Experimental Data, 2015 CEC*, 2015, Tucson AZ.
2. Miller, J., P. Bhandari, K. Novak, and J. Lyra, *MLI Blanket Effective Emittance Variance and its Effect on Spacecraft Propellant Line Thermal Control*, 46th ICES, Vienna, Austria.
3. Stultz, J., et al., *Test-Derived Effective Emittance for Cassini MLI Blankets and Heat Loss Characterization in the Vicinity of Seams*, AIAA 29th Thermophysics Conference, San Diego, CA, 1995.
4. Vanderlaan, M., et al., *Repeatability Measurements of Apparent Thermal Conductivity of Multilayer Insulation (MLI)*, CEC 2017, Madison, WI.
5. Moeini, E., et al., "Thermal performance evaluation of a fabricated multilayer insulation blanket and validity of Cunnington-Tien correlation for this MLI," *Cryogenics Journal*, 2013.
6. Smith, C., I. McKinley, P. Ramsey, and J. Rodriguez, *Performance of Multi-Layer Insulation for Spacecraft Instruments at Cryogenic Temperatures*, 46th ICES, Vienna, Austria.
7. Lin, E., and J. Stultz, *Cassini MLI Blankets High-Temperature Exposure Tests*, AIAA 33rd Aerospace Sciences Meeting, 1995.

7 Heat Pipes

Heat pipes are devices that transfer a large amount of heat by using the latent heat of vaporization of a working fluid to transport heat from one end to the other. They are a go-to for removing large amounts of heat with as little as possible mass and are essential for miniaturizing spacecraft. Packaging engineers who design ground-based electronics are well aware of these devices, and now they are making a strong showing in space electronics. As shown in Figure 7.1, the simple heat pipe has three main parts: the evaporator, condenser, and adiabatic section. The portion of the heat pipe that extracts heat is called the evaporator and like a refrigeration cycle, it absorbs heat by evaporating the working fluid, such as water. The water condenses from vapor back to liquid at the opposite end of the heat pipe, known as the condenser. The middle portion of the heat pipe that allows for vapor transport is called the adiabatic section. The water vapor then transports from one end of the heat pipe to the other by means of change in vapor pressure inside the heat pipe, which is lowest at the condenser due to the large change in specific volume as the vapor converts from vapor back to liquid, creating a slight suction that promotes the transport of vapor.

The wick inside the heat pipe aids the transport of liquid from the condenser back to the evaporator to repeat the cycle. In the heat pipe shown in Figure 7.1, the wick can be made of many things or can be machined grooves in the tube. For a sintered copper heat pipe, the wick is made of small, sintered copper beads that have a large capillary pressure and can pump fluid even against gravity. This allows it to operate in any orientation, thus aiding the testing of the hardware. It is a primary focus for low-cost, high-performance heat transfer for space applications.

As will be shown through this chapter, many kinds of heat pipes exist and target the many thermal requirements of spacecraft.

VARIABLE CONDUCTANCE HEAT PIPES (VCHPS)

VCHPs are designed to control the conductance of the working fluid by means of a non-condensable gas that restricts the flow of the working vapor. Liquid trap or diode heat pipes prevent the backwards flow of vapor by trapping the working fluid in its liquid form inside a bottle. VCHPs are typically used when power needs to be conserved during a survival condition, as they can easily be turned off and the only heat transfer occurs through the thin wall tube, which if sufficiently long has a small conductance. They also can turn themselves off when the environmental temperature quickly drops. They require a small heater at the reservoir bottle that contains the non-condensable gas, typically nitrogen.

DOI: 10.1201/9781003247005-7

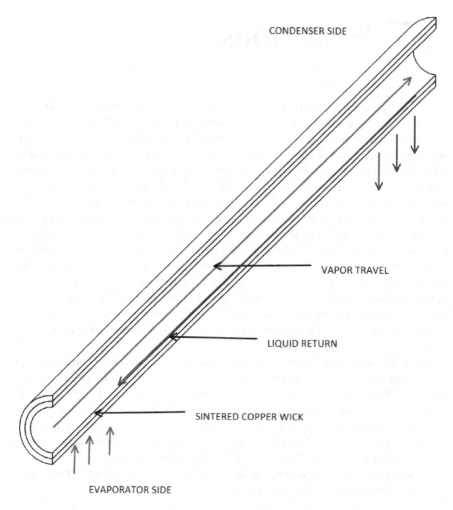

CONDENSER SIDE

VAPOR TRAVEL

LIQUID RETURN

SINTERED COPPER WICK

EVAPORATOR SIDE

FIGURE 7.1 Heat pipe operation. The water evaporates and travels up the heat pipe from the evaporator side to the condenser side, where it condenses and travels back to the evaporator.

CRYOGENIC SWITCHING HEAT PIPES (CSHPs)

CSHPs are designed to operate at extremely cold temperatures and like VCHPs can be turned off. The off mechanism happens by trapping the liquid in a separate bottle external to the heat pipe and reintroducing the liquid to turn the heat pipe back on. They can be arranged to work as thermal diodes, so heat travels in one direction and cannot travel back in the opposite direction. They are beneficial for uses on devices like focal plane arrays or cryogenic charged coupled device (CCD) that on occasion require an annealing process, in which the device is heated above around room temperature, then cooled down to cryogenic levels. They are also used when devices require de-icing. Without the ability to turn off the heat pipe, annealing and de-icing from cryogenic temperatures would require the entire radiator to be warmed up to room temperature, wasting a lot of power.

WATER SINTERED-COPPER HEAT PIPES

These heat pipes are in a category called constant conductance heat pipes (CCHP) and mostly used in terrestrial applications well above the freezing temperature of water. New Space, with their push for miniaturization, will drive the demand for these heat pipes, which have had a difficult time being adopted into the mainstream for space applications. These heat pipes when designed properly can be frequently frozen and thawed without concern of rupture.

NANO HEAT PIPES (NHPs)

NHPs are not nano-length heat pipes, but instead use a nanofluid to improve the overall performance of the heat pipe and in some cases remove the dependency on the wick to operate.

The next studies go into depth for each heat pipe technology and serve to give the reader a better understanding of how these heat pipes are implemented into real spacecraft applications.

THERMAL SWITCHING CRYOGENIC HEAT PIPE

Introduction. In the summer of 2009, JPL contacted ATK Aerospace Systems and described a need for a heat transport system (HTS) to link two CCD cameras, within the astrometric beam combiner (ABC) instrument on the SIM Lite Astrometric Observatory to a cryoradiator. Formerly the Space Interferometry Mission (SIM), SIM Lite will utilize optical interferometry to determine the positions/distances of stars much more accurately than has any previous program. The JPL HTS thermal requirements are as follows: (1) hot-side temperature of 150 K, (2) hot-side heat load of 6–12 W, (3) cold-side cryoradiator at 140 K, (4) transport length of 1.4 m, (5) periodic hot-side decontamination heating to 293 K with minimal heater power, (6) modest flight heritage, and (7) low cost/manufacturing complexity. This section describes the novel solution that was developed to meet the aforementioned requirements [1–3].

Concept. The solution that was developed to meet the HTS thermal requirements is a cryogenic heat pipe with thermal switching capability. This type of device is very similar to a cryogenic diode heat pipe (CDHP), which allows heat to flow only in one direction (forward mode). A common CDHP implementation involves positioning a cold-biased liquid trap (LT) on the evaporator end so that if the condenser becomes hot, the LT removes fluid from the heat pipe so as to prevent (condenser-to-evaporator) reverse mode heat flow. In a thermal switching cryogenic heat pipe, the LT is positioned on the condenser end and has its own cooling source. During normal operation, the small LT heater keeps the LT warm enough so that it is filled with vapor while the heat pipe is ON. To turn the heat pipe OFF, the LT heater is turned off and all the working fluid migrates to the LT. While the heat pipe is in the OFF condition,

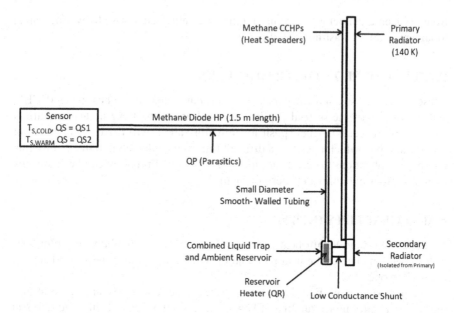

FIGURE 7.2 Concept for a thermal switching cryogenic heat pipe.

just a small amount of heater power is required on the evaporator end to achieve a significant (decontamination) temperature rise. The heat pipe can be turned back ON by simply repowering the LT heater. To develop the system described herein, thermal switching capability was implemented by appropriately modifying an Al axial groove heat pipe. Methane was used as the working fluid. The heat pipe modifications included the addition of (a) a small secondary radiator (SR) thermally isolated from the primary radiator (PR), (b) an LT plumbed to the cryogenic heat pipe with small-diameter tubing and thermally coupled to the SR with a low-conductance shunt (note: the LT is similar in design/implementation to the LTs used on the CRISM2 and the Three-Color Experiment methane CDHPs flying on the Mars Reconnaissance Orbiter [MRO] and an early Defense Support Program [DSP] satellite, respectively), (c) a small liquid trap heater, and (d) an ambient tank (AT) to reduce the fill pressure. Figure 7.2 illustrates the concept (AT omitted).

Trades. The important system design features that were traded to develop the solution that would best meet the requirements and the key issue(s) associated therewith are listed later. Brief comments on items 1–10 follow next. (1) Working fluid ethane vs. methane; (2) heat pipe architecture: axial groove vs. non-axial groove; (3) flight heritage: CRISM vs. three-color experiment; (4) fill pressure: ambient tank vs. high pressure heat pipe; (5) transport capacity: small diameter/margin vs. large diameter/ margin; (6) radiator sizing: small radiator/long test vs. large radiator/fast test; (7) shunt conductance: low G/Q, where G is shunt conductance, and Q is heat, long cooldown vs. high G/Q, short cooldown; (8) evaporator/condenser lengths: JPL lengths of 15 cm/71 cm vs. longer or shorter; (9) parasitics: simulation in test heaters vs. temperature-controlled shroud; and (10) decontamination: liquid trap vs. no liquid trap. With regard to the working fluid, methane is the superior choice as its liquid transport factor (NT = $\sigma\Delta H/n$) is roughly equal to that of ethane at 140 K, but methane experiences a steep fall-off in transport capacity above 140 K, becoming zero at its critical temperature of 191 K; thus, decontamination power is lower with methane. With regard to the heat pipe architecture, the axial groove design is the simplest, least costly, highest test readiness level (TRL) approach. With regard to flight heritage, features of both CRISM and the three-color experiment CDHP systems were combined, ensuring the selected approach would have significant flight heritage. With regard to fill pressure, using an ambient tank reduces the fill pressure by five times or more, improving system durability, reliability, safety, and flight qualifiability. With regard to transport capacity, a heat pipe with a larger diameter than necessary provides design margin if transport requirements grow. With regard to radiator sizing (for testing), using a larger radiator than one used for flight expedites characterization testing and modestly reduces development cost. With regard to shunt conductance (G), the lower OFF-state heater power of a low-G shunt arguably outweighs the faster cooldown rate afforded by a high-G shunt. With regard to evaporator and condenser lengths, the initial design lengths selected by JPL provide acceptable margin on the HTS thermal requirements. With regard to parasitics simulation (during testing), the simplicity of using heaters outweighs the flight system accuracy of a temperature-controlled shroud. Last, with regard to (ease of) decontamination, the lower heater power and thermal switching efficiency of having an LT greatly outweighs the design simplicity of not having one.

Design. Figure 7.3 illustrates the design features of the thermal switching cryogenic heat pipe. The key design features are as follows: (a) heat pipe extrusion (6063 Al, axial groove, 1.5 cm OD, 10 cm ID, 75 W-m capacity with 140 K methane);

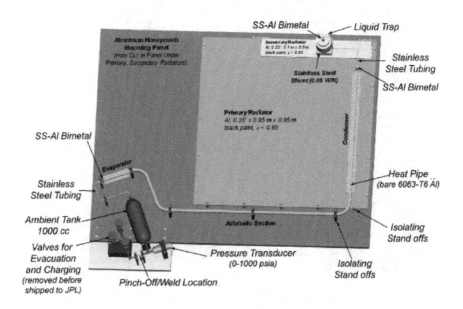

FIGURE 7.3 Design of the thermal switching cryogenic heat pipe.

(b) evaporator (15 cm × 5 cm flange); (c) adiabatic section (1.4 m length); (d) condenser (71 cm × 5 cm flange); (e) ambient tank (stainless steel, 1000 cc), (f) primary radiator (6061 Al, 0.95 m × 0.95 m × 0.6 cm, black paint on both sides, isolated from Al honeycomb panel with Delrin rods); (g) secondary radiator (6061 Al, 0.1 m × 0.6 m × 0.6 cm, black paint on both sides, isolated from primary radiator by Delrin pins); (h) heat pipe supports (Delrin isolators, Al honeycomb panel with cutout for radiators, Delrin isolators underneath panel); (i) pressure transducer (0–6.9 MPa); (j) valves (for fill/vent, removed before shipment to JPL); (k) liquid trap (6063 Al, CRISM LT design); (l) LT shunt (stainless steel, 0.05 W/K); and (m) working fluid (99.999% methane, 4.1 MPa fill pressure).

TESTING

After the thermal switching cryogenic heat pipe was manufactured and assembled in accordance with the Figure 7.3 design at the ATK manufacturing facility in Beltsville, MD, the unit was then configured for thermal vacuum testing. Figure 7.4 illustrates the overall test setup, thermocouple locations, heater placement, and MLI coverage (note: heater labels are preceded by the letter "H" and MLI is depicted by the dotted red lines). The unit was mounted in ATK Chamber E, which has a full 360° LN2-cooled, box-shaped shroud. Figure 7.5 illustrates the thermal vacuum

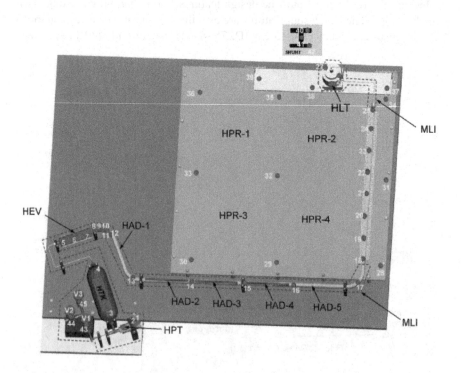

FIGURE 7.4 Test setup for the thermal switching cryogenic heat pipe.

FIGURE 7.5 ATK thermal vacuum chamber E. The initial test plan called for three primary performances.

chamber used at the ATK test facilities in Beltsville, MD. After the placement of the thermal switching cryogenic heat pipe assembly on the flat interior bottom surface of the shroud, a transit and ruler were used to level the unit so that it would have a slightly adverse tilt. The arbitrary tilt specification specified in the test plan required that the evaporator would be 1.3 +/– 0.5 mm above the condenser.

Characterization/acceptance tests: (1) Test 1: Cooldown/ON-OFF-ON (during which the heat pipe would be cooled down from ambient temperature to an ON condition, turned OFF with the LT, then turned back ON); (2) Test 2: Normal Operation (during which evaporator and parasitic heater powers up to the maximum HTS requirement); and (3) Test 3: Decontamination (during which three different decontamination heating options would be utilized). With regard to Test 3, the three decontamination heating options consisted of the following: (i) Decontamination Option 1: high power (120 W) heating of the primary radiator to 191 K plus a small amount of evaporator heater power; (ii) Decontamination Option 2: modest power (40 W) heating of the evaporator so that it would deprime and the transport capacity of the heat pipe would fall as the primary radiator temperature rose; and (iii) Decontamination Option 3: low power (3–5 W) implementation of the LT as explained in the earlier part of the paper plus a small amount of evaporator heater power.

Table 7.1 lists the original test matrix. Due to project schedule constraints, JPL decided that only tests 1–1, 1–2, 2–6, and 3–11 would be carried out (unless opportunities arose during testing to carry out additional tests without adversely impacting project schedule). In addition, due to a slower than expected cooldown rate of the LT during testing (which occurred due to an analysis oversight wherein the transient cooldown of the thermally shunted LT was not included in pretest thermal predictions), the abridged test program indicated by Table 7.1 was further modified. The actual tests that were carried out over a 31.75-hour period beginning at 6 am on 4/24/10 are listed next. As indicated, Test 1–2 was eliminated, and modified versions of Test 3–13 and Test 3–7 as well as a repeat of Test 2–6, which is denoted below as Test 2–6r, were carried out. The temperature and methane pressure time-histories corresponding to each of the tests listed next are provided in Figures 7.6–7.11. Test 1–1 ($t = 0.00$–10.25 hours; test duration 10.25 hours); Test 2–6 ($t = 10.25$–11.25 hours;

TABLE 7.1
Test Matrix (Original Unabridged Plan)

Test #	Steady State #	Description	QEVAP(W)	QCOND (W)	QLTRP (W)	QPARA (W)	Time (Hrs)
1	1	Initial Cooldown	0 ramp to 8	Q(TPR = 140 K)	2	0	10+1=11
	2	Turn-Off	Q (TEVAP = 293 K) ~ 3	Q(TPR = 145 K)	0	0	3+1 = 4
	3*	Turn-On	3 ramp to 8	Q(TPR = 140 K)	2	0	3+1 = 4
2	4*	Normal Oper. Min, no QP	6	Q(TPR = 140 K)	2	0	1+1 = 2
	5*	Normal Oper. Max, no QP	12	Q(TPR = 140 K)	2	0	1+1 = 2
	6	Normal Oper. Max, QP	12	Q(TPR = 140 K)	2	2.5	1+1 = 2
3	7*	Decontam. Option 1	Q (TEVAP = 293 K) ~ 3	Q(TPR = 191 K)	2	2.5	3+1 = 4
	8*	Normal Oper. Max, QP	12	Q(TPR = 140 K)	2	2.5	3+1 = 4
	9*	Decontam. Option 2	Q (TEVAP = 293 K) ~ 40	0	2	2.5	3+1 = 4
	10*	Normal Oper. Max, QP	12	Q(TPR = 140 K)	2	2.5	3+1 = 4
	11	Decontam. Option 3	Q (TEVAP = 293 K) ~ 3	Q(TPR =100 K)	0	2.5	3+1 = 4
	12*	Low Temp. Operation 1	Q (TEVAP = 110 K)	0	2	0	3+1=4
	13*	Low Temp. Operation 2	Q (TEVAP = 120 K)	0	2	0	3+1 = 4
		Test Completion				TOTAL	53

* Optional tests per JPL direction

test duration 1.00 hours); Test 3–13 ($t = 11.25$–15.00 hours; test duration 3.75 hours); Test 3–7 ($t = 15.00$–18.75 hours; test duration 3.75 hours); Test 3–11 ($t = 18.75$–29.75 hours; test duration 11.00 hours); Test 2–6r ($t = 29.75$–31.75 hours; test duration 2.00 hours). Initial cooldown and turning ON the thermal switching cryogenic heat pipe is illustrated in Figures 7.6a and 7.6b. Figure 7.6a is a zoomed-out view to show the variation in methane pressure from the fill pressure value of 4.1 MPa (600 psi) to the operational value of 1.0 MPa (145 psi) when temperatures had steadied out. Figure 7.6b is a zoomed-in view to better illustrate the heat pipe temperatures. At the conclusion of Test 1–1, with an evaporator heater power of 10 W versus the planned 8 W, the thermal switching cryogenic heat pipe had started up successfully.

Two anomalies were observed during Test 1–1 that should be mentioned before proceeding. First, due to the closeness of the evaporator and adiabatic temperatures, it was hypothesized that thermocouples 5–7 may have been placed on the unheated side of the evaporator. After reviewing the assembly photos, that hypothesis was

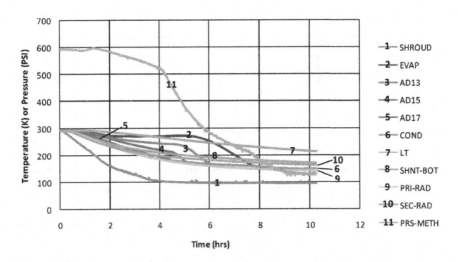

FIGURE 7.6a Initial cooldown and turn ON: Test 1–1 zoomed-out view

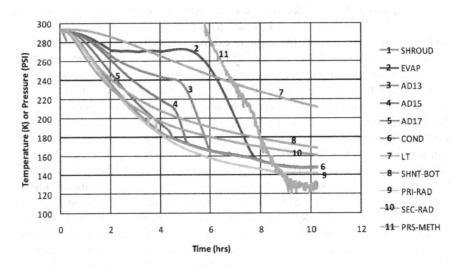

FIGURE 7.6b Initial cooldown and turn ON: Test 1–1 zoomed-in view

confirmed. Thus, in the remainder of the figures presented herein, the evaporator temperatures will be slightly lower than they should be. The likely temperature underestimation, considering only evaporative film resistance, should be roughly 3 K for a 12 W heat load. Second, the resistance of the evaporator heaters (five small heaters wired in parallel) at room temperature was about 4 ohms. However, when the system had cooled to 150 K, the resistance was only about 2 ohms. This resistance change was verified by the data acquisition equipment and by hand measurements. The cause of this change (decrease) is unknown, but it might be due to either coefficient of

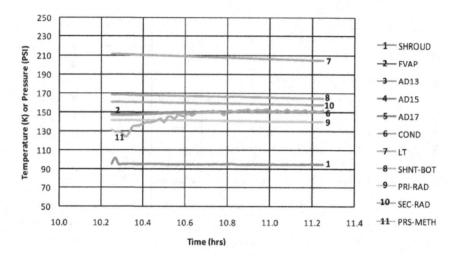

FIGURE 7.7 Normal operation: Test 2–6.

thermal expansion (CTE) effects (increased clamping force at the junction where the five evaporator heaters were wired in parallel) or the presence of ultra-high-purity copper traces in the heaters (which may exhibit temperature-dependent electrical resistance). It is assumed that the data acquisition equipment was correct and that there was not simply an error in measuring the heater resistance and the corresponding evaporator heater power. Figure 7.7 illustrates the results of Test 2–6, the normal operation test. Overall, Test 2–6 demonstrates successful normal operation of the thermal switching cryogenic heat pipe at a primary radiator temperature of 140 K, a condenser temperature of 150 K, an evaporator temperature of 151 K, a maximum evaporator heat load of 12 W, and a parasitic heat load (spread evenly over the adiabatic section of the heat pipe) of 2.5 W. The curve at the top of Figure 7.6 is the LT. Based on the slow LT cooldown, a decision was made to deviate from the (abridged) test plan and proceed to Test 3–13. Figure 7.8 illustrates the results of Test 3–13, the low operating temperature test. As indicated, the low operating temperature target of 120 K indicated in Table 7.1 was modified to 135 K so that enough time would be available to carry out Test 3–7, decontamination option 1. Overall, Test 3–13 demonstrates successful low temperature operation of the thermal switching cryogenic heat pipe at a primary radiator temperature of 127 K, a condenser temperature of 135 K, an evaporator temperature of 135.5 K, and heat loads identical to Test 2–6.

Figure 7.9 illustrates the results of Test 3–7, the decontamination option 1 test. The planned test procedure was to heat the condenser to 191 K to deactivate the heat pipe. To implement this procedure, the LT needed to be heated along with the condenser to prevent liquid trapping. However, given the slow cooling rate of the LT, doing so would surely have added test time and jeopardized project schedule. So, decontamination option 1 was modified as explained later. Initially, 150 W was applied to the primary radiator at just after 15 hrs. At just after 17 hrs, the 175 K condenser temperature had just risen above the LT temperature. Shortly thereafter, well before the

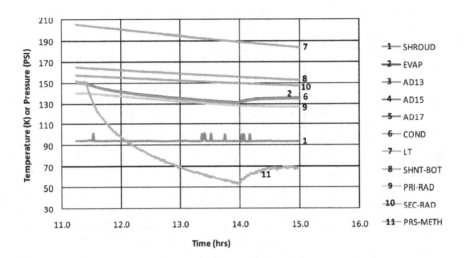

FIGURE 7.8 Low temperature operation: Test 3–13.

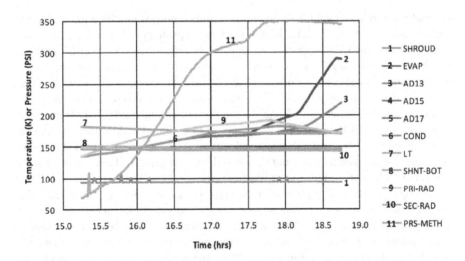

FIGURE 7.9 Decontamination Option 1: Test 3–7

condenser was at 191 K, the evaporator deprimed, as the LT had begun trapping fluid. At 17.75 hrs, the primary radiator heater was turned off. At 18.25 hrs, evaporator heater power was increased to 26 W. Shortly thereafter, at 18.75 hrs, the evaporator was at 293 K. This test thus demonstrated a new hybrid decontamination procedure that would be utilized to carry out Test 3–11 as described later. Figure 7.10 illustrates the results of Test 3–11, the decontamination option 3 test (note: the roughly 10 hr delay in conducting this test following the previous test was necessary to wait for the LT to cool down to the required temperature). Based on the earlier discussion, the

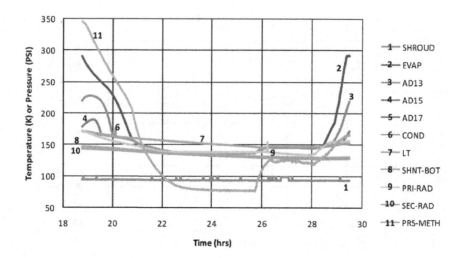

FIGURE 7.10 Decontamination Option 3: Test 3–11

originally planned decontamination option 3 procedure was modified to the following four steps: (1) turn off the LT heater; (2) heat the primary radiator so that the condenser is above the LT temperature (note: primary radiator heater power is only that
which is necessary to raise the condenser temperature to, or slightly above, the LT
temperature); (3) heat the evaporator to rapidly raise its temperature to 293 K (note:
if heat-up time is not of paramount importance as it was in this test, lower evaporator heater powers, resulting in longer heat-up times, can be used); and (4) once the
evaporator is at 293 K, reduce the heater power to maintain temperature. As can be
seen in Figure 7.10, this procedure worked very well, and it is recommended that this
procedure be utilized to carry out decontamination heating in future ground tests
and in the flight system. Figure 7.11 illustrates the results of Test 2–6r, a repeat of the
normal operation test. The purpose of this test was to demonstrate how quickly the
system returns to normal operation after a hybrid decontamination procedure. One
advantage of having an LT is that decontamination can be carried out without appre-
ciably altering system pressure. As seen in Figure 7.11, the methane pressure changes
by only 0.3 MPa (40 psi) in transitioning from the decontamination condition to the
normal operating condition. Without an LT, the pressure variation and time to return
to normal operation would have been greater. The test closest to simulating a non-LT
system is Test 3–7 (see Figure 7.9). As seen, the high 2.4 MPa (350 psi) peak pressure
during decontamination about doubles the time necessary to return to normal operation (see Figure 7.10).

Conclusion. The two main conclusions of this effort are as follows: (1) the thermal switching cryogenic heat pipe functions very well as a heat transport system
and easily meets all JPL heat transport, operating temperature, and temperature stability requirements (note: the +/–2 K/hr temperature stability requirement was not
expressly addressed in this section because it was so easily achievable), and (2) the
thermal switching cryogenic heat pipe functions very well as a thermal switch by

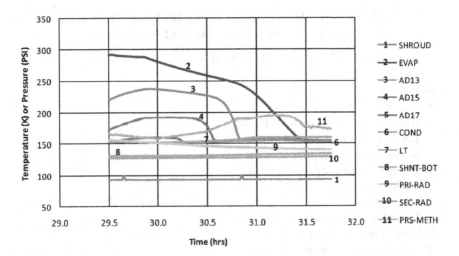

FIGURE 7.11 Normal Operation Repeat: Test 2–6r

utilizing the working fluid capturing capability of the LT. It is recommended that the hybrid decontamination procedure developed and demonstrated during the test program be utilized in future ground testing and in the flight system. The hybrid decontamination procedure, which combines primary radiator heating, LT fluid capture, and evaporator heating, is a very fast and efficient way to turn the thermal switching cryogenic heat pipe OFF and heat the evaporator to 293 K for CCD decontamination. Turning the thermal switching cryogenic heat pipe back ON from the OFF state is a very fast and efficient process as well.

Acknowledgement. The work performed herein was carried out at the ATK Aerospace Systems facility located in Beltsville, MD, under contract to the NASA Jet Propulsion Laboratory (JPL). The authors would like to acknowledge the following ATK employees for their valuable contributions during the program (in alphabetical order): Greg Davis, James Govern, Joe Hudson, Mike Humphrey, Jessica Kester, Donna King, Chris Smith, Ben Weetman, Stuart Whitelock, and James Yun.

REFERENCES

1. Goullioud, R., J. Catanzarite, G. Dekens, M. Shao, and J. Marr, "Overview of the SIM Lite PlanetQuest light mission concept," *Proceedings of the SPIE 7013*, Pasadena, CA, 2008, pp. 7013–177.
2. Bugby, D., J. Garzon, M. Beres, C. Stouffer, D. Mehoke, and M. Wirzburger, "Cryogenic diode heat pipe system for cryocooler redundancy," *Proceedings of the SPIE 5877*, Tucson, AZ, 2005, pp. 321–329.
3. Misselhorn, J., P. Brennan, and G. Fleischman, "Application of heat pipes to a cryogenic focal plane space experiment," *AIAA-87–1648, 22nd AIAA Thermophysics Conference*, Honolulu, HI, 1987.

FREEZE-THAW ANALYSIS OF SINTERED COPPER WATER HEAT PIPES FOR SPACE APPLICATIONS

Juan Cepeda-Rizo, [1] Jeremiah Gayle, [2] David A. Edwards, [3] and John G. Thayer, [4]

Nomenclature

A: area, units L2

B: arbitrary constant, variously defined

c: specific heat, units $L^2 = T^2\Theta$

D: thermal diffusivity, units $L^2 = T$

d: diameter of pore or particle, units L

F: trapped water ratio

f: charging factor

k: thermal conductivity, units $ML = T^3$

L: latent heat of fusion, units $L^2 = T^2$

R: radial measurement, units L

r: radius of pore or particle, units L

St: Stefan number

s(t): freezing boundary, units L

Θ: temperature, units T

t: time, units T

x: distance along pore, units L

y: distance across pore or copper, units L

Vr: volumetric expansion factor

α: linear expansion factor

ζ: similarity variable

$\Theta(\zeta)$: similarity solution for temperature, units

ρ density, units $M = L3$

ϕ: void fraction

Other Notation

c: as a subscript, used to indicate copper

h: as a subscript on R, used to indicate the heat pipe

i: as a subscript, used to indicate ice

m: as a subscript on R, used to indicate the wick

p: as a subscript on d, used to indicate a pore

v: as a subscript on R, used to indicate the inner wick radius

w: as a subscript, used to indicate water

0: as a subscript, used to indicate an initial value

I. INTRODUCTION

Heat pipes use the heat capacity of a working fluid to dissipate heat. In the case we analyze, these devices consist of a copper pipe (sealed at both ends) with an annular sintered copper wick inside and contain water as the working fluid [1–10]. Capillarity forces induced by the porosity of the wick enhance the circulation of water in the pipe. The core of the pipe is then set to near vacuum. Liquid water at the hot end is vaporized, absorbing heat. Given that the pipe is sealed, the water vapor travels to the other (cold) end, where it condenses in the wick, which is adjacent to the cold environment of space. The water then travels through the wick to the hot end, where the process repeats. These designs have been used extensively in terrestrial applications (cooling computer chips, etc.), but in those cases the cold end has not been held at temperatures which would freeze the water inside. Therefore, care must be taken to ensure that the pipe will not fail in such an environment. To do that, experimentalists run a series of freeze-thaw cycles to determine whether the repeated freezing and thawing of the water inside will damage the copper components.

[1] Thermal Systems Engineer, NASA/Jet Propulsion Laboratory, 4800 Oak Grove Dr., Pasadena, CA 91109, M/S 103–116

[2] Mechanical Engineer, NASA/Jet Propulsion Laboratory, 4800 Oak Grove Dr., Pasadena, CA 91109, M/S 125–123.

[3] Professor, Department of Mathematical Sciences, 511 Ewing Hall, University of Delaware, Newark, DE 19716-2553

[4] John Thayer, Group Leader, Aavid Thermacore, Aero/Active Systems Engineering, Lancaster, PA

The thermophysics of the sintered copper heat pipe must be understood such that its behavior is predictable and repeatable for a given set of conditions to be considered ready for flight. The qualification program used to verify this understanding must expose the units to relevant environments that bound the flight scenarios. It must also include data collected that sufficiently corroborates the understanding.

There are two modes of failure that we wish to investigate. In the first, water accumulates in the cold end and then freezes. The resulting expansion of the liquid water is enough to burst the pipe wall and has been seen experimentally [1]. We call this wall failure. The sintered copper wick forms a porous medium with a small pore size, which enhances liquid transport from the cold end to the hot end. In the second type of failure, water in the wick freezes and expands, deforming the porous media structure. With a larger pore size, the speed of transport in the wick decreases, reducing the heat pipe's effectiveness. We call this wick failure. In this section we will examine both types of failures, with an eye toward establishing tolerances below which we expect the wick to remain functional.

Freeze-out and freeze-thaw tests were created to mitigate risk. Copper water heat pipes were embedded in an aluminum chassis. A heat load was applied while the condenser end was chilled in a cycle from +20 °C to −25 °C and back again to +20 °C. The heat pipe did encounter freeze-out. The evaporator dried out when the condenser was held at −25 °C. That dry-out persisted when the condenser thawed out, but recovered

FIGURE 7.12 Cross-section of a sintered copper heat pipe [2]

when the heat load was power cycled. The effect was minimal in gravity-neutral conditions and 50 °C condenser temperature, near the upper end of the operating temperature against damage due to freezing of the heat pipe. The freeze-out test was originally designed as a mitigation, in which the heat pipes would undergo freezing and "dry-out," a situation in which the pipe is forced to freeze at one end, while the other end experiences enough power to dry out the wick of the heat pipe.

A thermal vacuum test anticipated bringing the heat pipes down below freezing while the instruments were powered on. The freeze-out test was performed on six heat pipes for three cycles as a way to have a 95% confidence that the heat pipes would function properly. The freeze-thaw test froze and thawed each heat pipe that was ordered as a way of verifying that the heat pipe could survive unpowered freezing; this was done for 100 cycles for every heat pipe. An internal peer review board discussed that the freeze-thaw test was the best test for demonstrating that the heat pipes could survive repetitive freezing and thawing. Additional unanticipated concerns with freezing of heat pipes came up during the testing and showed concerns with adverse orientation, causing a hysteresis in performance. However, the tests have shown that heat pipe reprimes with no hysteresis in the gravity-neutral or zero-gravity condition.

II. ANALYSIS OF HEAT PIPE FAILURE

A. GEOMETRIC CONSIDERATION OF THE WICK INSIDE THE TUBE

The description of the heat pipe and its radial dimensions are shown in Figure 7.12. We begin by examining the case of pipe failure. We introduce a simple one-dimensional model. We scale the width of the wick to be 1, and just consider the proportion of each of the three components in the cross-section.

At roughly room temperature (T_{w0} = 20 °C, where the subscript "w" refers to "water," the void fraction of the porous medium is given by $(0 \leq \phi \leq 1)$. Hence the proportion of copper in the wick is given by $1 - \phi$. Then the voids are filled with water to some percentage f (called the charging factor), so the fraction of water in the cross-section is given by ϕf.

Next, we bring the temperature of the pipe down to T_{i0} = 0 °C, where the subscript "i" refers to "ice." We choose this value since this is the where the density of water is lowest, which corresponds to the largest expansion. The copper will

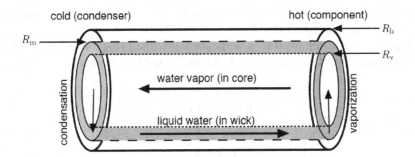

FIGURE 7.13 Description of a heat pipe's inner workings.

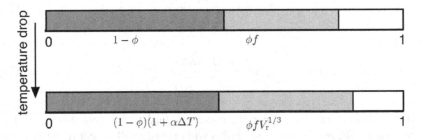

FIGURE 7.14 One-dimensional model of expansion.

contract by a factor of $1 + \alpha \Delta T$, where α is the linear thermal expansion coefficient, and $\Delta T = T_{i0} - T_{w0}$. α is roughly constant for a metal like copper. Note also that $\Delta TT = -20\ °C < 0$, so the metal does indeed contract. On the other hand, the water will expand. The coefficient of thermal expansion for water is not constant, so we use specific values of the density of water. In particular, we have that the volume of water will expand by the following volumetric expansion factor:

$$V_r = \frac{\rho_i}{\rho_w} \tag{7.1}$$

We note that $V_r > 1$ since the density of water is higher at room temperature than at freezing. To convert this to a linear expansion factor, we simply take the cube root. Hence, we have that at freezing, the amount of space taken up by the copper and the water is given by:

$$\underbrace{(1-\phi)(1+\alpha \Delta T)}_{copper} + \underbrace{\phi f V_r^{1/3}}_{water} \tag{7.2}$$

We say that the pipe has failed if the Expression (7.2) is greater than 1. Hence, the ice has expanded enough that it will burst the pipe wall or expand into the inner core.

Once in the core, the water can pool during thaw cycles, eventually freezing and bursting the pipe during a later cycle.

Typically, ϕ is determined for a particular wick, and f is at the discretion of the manufacturer. Therefore, we construct an upper bound for f given a particular ϕ :

$$(1-\phi)(1+\alpha\Delta T)+\phi fV_r^{\frac{1}{3}} \le 1 \tag{7.3}$$

$$\phi fV_r^{1/3} \le 1-(1-\phi)(1+\alpha\Delta T) \tag{7.4}$$

$$f \le \frac{-\alpha\Delta T + \phi(1+\alpha\Delta T)}{\phi V_r^{\frac{1}{3}}} \tag{7.5}$$

$$\le \frac{1+\alpha\Delta T}{V_r^{\frac{1}{3}}} - \frac{\alpha\Delta T}{\phi V_r^{\frac{1}{3}}}. \tag{7.6}$$

If $\phi = 1$ and there is no copper, the maximum loading percentage is $V_r^{-1/3}$. This corresponds to the percentage of water that will expand to fill the entire volume upon freezing. If $\phi = 0$ and there is no water, there is no maximum loading percentage.

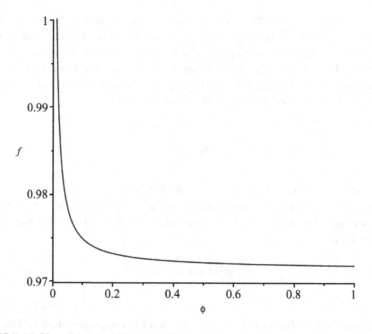

FIGURE 7.15 Plot of charging factor vs. void fraction.

A graph for the values in Appendix A is given in Figure 7.15 the upper limit for the charging factor is always over 97%. How then to explain the failure in [3]? Recall that in terrestrial applications, freezing is not an issue, so the charging factor can be 1, or even higher, if additional water is allowed to pool in the core. Hence, a heat pipe designed for terrestrial uses could undergo pipe failure upon freezing. Fortunately, heat pipes designed for use in space are usually charged to only about $f < 0.95$, which is under the upper bound in [3].

We conclude by noting that the failure of the heat pipe is quite different from the failure of copper water pipes in houses.

In a house water pipe, the cold pipe freezes the water from the outside in, forming an ice annulus. As the annulus solidifies into a solid plug, the ice then begins to expand parallel to the pipe. Expansion toward the house increases pressure on the water in the pipe downstream of the plug. It is this pressure that then bursts the pipe. In contrast, the heat pipe is near vacuum. So, the ice itself must burst the copper pipe. Moreover, this phenomenon is quite different in terms of geometry. Most engineering discussions of pipe bursting focus on the case where the pipe wall thickness is small compared to the pipe diameter. But we see from Appendix A that the pipe wall thickness is comparable to the pipe diameter, which is a regime not generally discussed.

B. WICK FAILURE

We next examine the case of wick failure. We wish to see if the thermal expansion of the water as it freezes will deform the structure of the pores in the wick. If the pores become too large, that may lessen the wicking effect and cause the heat pipe to fail. (To our knowledge, this has not been observed experimentally.)

As the porous media itself is made of copper, it will conduct heat quickly. Hence, we expect the ice to move outward from the copper toward the center of the pores. In general, this will cause a pressure gradient which will pump water through the porous media (see left of Figure 7.16). Since the inner core is near vacuum, water will tend to flow from the outer portions of the pipe to the empty inner portion of the wick. Then that water will also come into contact with the cold copper wick, where it will freeze.

However, as the ice from neighboring beads comes together, they can isolate a water pocket (small white area in center of right of Figure 7.17). As this pocket freezes, it will exert pressure on the neighboring ice. It is our belief that if these pockets are small enough, they will not cause enough pressure to distort the wick matrix, which has of course been trapped in ice from the beginning.

We now present a problem to illustrate our analysis. But first, some caveats:

1. The idea of isolating the copper beads in the water is not realistic. We could think of them being connected by thin sinter connections made of copper, or consider the microstructure more carefully.
2. In the right of Figure 7.17, the ice region is shown as a series of overlapping circles, and we will continue that model here. However, in reality surface energy considerations would cause the boundaries of the water pocket to be convex.

3. It is much easier to trap water in two dimensions than in three dimensions. In particular, consider the problem of circle packing in two dimensions. This will cause many inclusions, as described later. But the problem of sphere packing does not. In particular, a perfect sphere packing still has a continuous pore structure through it. Hence, there will have to be more overlap (as in the right of Figure 7.16) before the inclusions are closed. This will reduce their surface area.

In the light of the aforementioned, the analysis we present will give a conservative estimate of the volume of inclusions—most assuredly, too conservative.

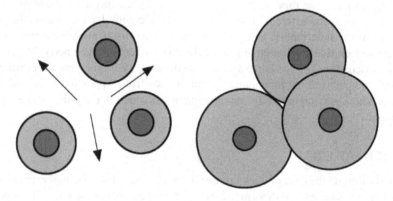

FIGURE 7.16 Schematic of pumping action caused by freezing. Dark gray circles correspond to copper; light gray circles correspond to ice. Left: illustration of pumping process at beginning of freezing. Water flows in direction of arrows. Right: As freezing continues, a small quantity of water is trapped by the overlapping ice circles.

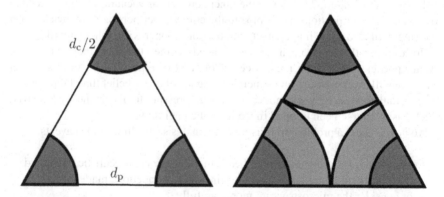

FIGURE 7.17 Schematic of trapping water by freezing of triangular lattice. Triangle is repeated in six-fold symmetry in lattice. Dark gray regions correspond to copper, light gray regions correspond to ice, and white regions correspond to water. Left: array before freezing. Right: trapping of water in matrix.

We begin by considering a triangle (really part of a hexagonal array), as shown in Figure 7.17 The triangle is equilateral with side length $d_p + d_c$, where dp is the diameter of the pore and d_c is the diameter of the bead (the subscript "c" stands for "copper"). Hence, the area of the triangle is given by:

$$A_\Delta = \frac{\left(d_p + d_c\right)^2 \sqrt{3}}{4} \tag{7.7}$$

Taken together, the sectors form a semicircle of diameter d_c with area:

$$A_{c,\Delta} = \frac{\pi}{2}\left(\frac{d_c}{2}\right)^2 = \frac{\pi d_c^2}{8} \tag{7.8}$$

After freezing, the sectors form a semicircle of diameter $d_c + d_p$ with area:

$$A_{i,\Delta} = \frac{\pi\left(d_c + d_p\right)^2}{8} \tag{7.9}$$

$$F_\Delta = \frac{A_\Delta - A_{i,\Delta}}{A_\Delta - A_{c,\Delta}} = \frac{\left[\left(d_p + d_c\right)^2 \sqrt{3}/4\right] - \pi\left(d_c + d_p\right)^2/8}{\left[\left(d_p + d_c\right)^2 \sqrt{3}/4\right] - \pi d_c^2/8} = \frac{\left(d_p + d_c\right)^2\left(2\sqrt{3} - \pi\right)}{2\left(d_p + d_c\right)^2 \sqrt{3} - \pi d_c^2}$$

$$\tag{7.10}$$

With the values of the parameters from Appendix A, we have:

$$F_\Delta = 0.130$$

Note:

1. Because of the symmetry involved with the equilateral triangles, this is actually the largest percentage of water that can be trapped. Random placement greatly reduces the size of the inclusions (see right of Figure 7.16).
2. The analysis can be continued by hand (at least) for isosceles triangles. However, the analysis is complicated, must be broken into separate cases, and is not particularly illustrative.

C. 1D HEAT TRANSFER MODELS

In contrast to the description in Section 1 a more common model of porous media is to have regular geometric shapes (such as a network of cylinders) for the pores, rather than the media. This is motivated by the microstructure of the wick. In that case, water-filled cylinders closing up with ice would be equivalent to a pore clogging.

This sort of analysis could give vastly different results about whether all the water could be squeezed out of the medium.

In particular, consider the scenario illustrated in Figure 7.18. Consider two pores, A and B. Suppose that freezing proceeds linearly along the pores. Since A is more tortuous than B, ice from B will block A, trapping the water in the remaining length of A. If this water freezes in place, it could deform the porous medium adjacent to it.

We examine some simple models to get some estimates for the time scales of the heat transfer effects. First, we consider a pore as a single cylinder that penetrates the entire wick. Its diameter would be $d_p = 50\mu m$, which is much less than the thickness of the wick, which is 7.5 mm. Thus, the cylinder can be approximated by a one-dimensional model.

We wish to estimate the amount of time it would take for the water in the pore to freeze. At first, we consider a simple example where the water is cooled only from the outer surface of the wick. This leads to consideration of the one-phase Stefan problem where water freezes into ice. The details of the derivation may be found in [3]; we outline the steps as follows.

Let x be distance along the pore as measured from the outer wall of the wick (see Figure 7.19). We consider a semi-infinite problem; this is a reasonable simplification since we are trying to get only an estimate of the freezing front progression. $T_i(x,t)$, the temperature of the ice, follows the standard heat equation:

$$\frac{\partial T_i}{\partial t} = D_i \frac{\partial^2 T_i}{\partial x^2}, \quad D_i = \frac{k}{\rho c_i} \tag{7.11}$$

where c_i is the specific heat and k_i is the thermal conductivity of the ice. D_i is called the thermal diffusivity. Note that we have not put a subscript on the density, even though we know there is a drastic change in the density as the water freezes. The Stefan problem can be formulated with a change in density, though the setup is more complicated. Hence, we proceed with a constant density (namely the one for water at 0.01 °C) to follow the standard Stefan problem analysis. The domain of (1) is $0 < x < s(t)$, where $s(t)$ is the freezing boundary. At that boundary, we have:

$$T_i(s(t),t) = 0 \tag{7.12a}$$

$$k_i \frac{\partial T_i}{\partial x}(s(t),t) = \rho L \frac{ds}{dt} \tag{7.12b}$$

where L is the latent heat of fusion (freezing) of water. Equation (7.12b) says that the net heat flux into the freezing front will be used up in freezing the ice into its crystalline shape. Note that in the one-phase Stefan problem, the temperature in the water is assumed to be held at exactly 0. Hence, there is no contribution from heat flux in the water to (7.12b). We close the system with a boundary condition on T_i and an initial condition on s:

$$T_i(0,t) = T_{i0,} \qquad (7.13a)$$

$$s(0) = 0 \qquad (7.13b)$$

With these conditions, (7.11) has a similarity solution of the form:

$$T_i(x,t) = \Theta_i(\zeta),\ \zeta = \frac{x}{2\sqrt{D_i t}},\ s(t) = 2s_o\sqrt{D_i t} \qquad (7.14)$$

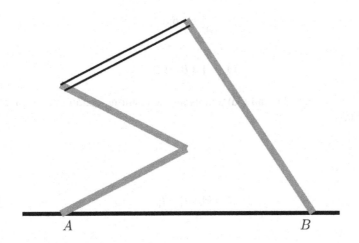

FIGURE 7.18 Pore structure: Gray—ice; white—water.

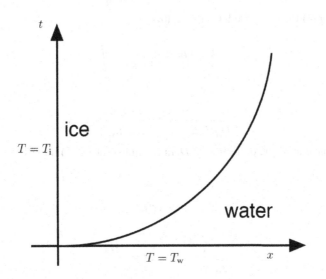

FIGURE 7.19 Schematic of Stefan problem.

Note that (4) automatically satisfies (3b). Substituting (4) into (1) and solving, we have:

$$-\frac{x}{4\sqrt{D_i t^3}}\frac{d\Theta_i}{d\zeta} = D_i \frac{1}{4D_i t}\frac{d^2\Theta_i}{d\zeta^2} \tag{7.15a}$$

$$-2\zeta\frac{d\Theta_i}{d\zeta} = \frac{d^2\Theta_i}{d\zeta^2}$$

$$\frac{d\Theta_i}{d\zeta} = B_1 e^{-\zeta^2}$$

$$\Theta_i(\zeta) = B_1 \operatorname{erf}\zeta + B_2 \tag{7.15b}$$

Substituting (4) into (2a) and (3a) and using the resulting conditions to solve for the B_i, we obtain:

$$\Theta_i(s_0) = 0 \tag{7.16a}$$

$$\Theta_i(0) = T_{i0} \tag{7.16b}$$

$$\Theta_i(\zeta) = T_{i0}\left(1 - \frac{\operatorname{erf}\zeta}{\operatorname{erf}s_0}\right). \tag{7.17}$$

Substituting (4) into (2b), we have the following:

$$\frac{k_i}{2\sqrt{D_i t}}\frac{d\Theta}{d\zeta}(s_0) = \rho L s_0 \sqrt{\frac{D_i}{t}} \tag{7.18}$$

$$\frac{k_i}{2D_i \rho L}\frac{d\Theta}{d\zeta}(s_0) = \frac{c_i}{2L}\frac{d\Theta}{d\zeta}(s_0) = s_0 \tag{7.19}$$

where we have used the definition of D_i in (1). Substituting (7) into (8), we obtain:

$$\frac{c_i}{2L}\left(-\frac{2T_{i0}}{\sqrt{\pi}}\frac{e^{-s_0^2}}{\operatorname{erf}s_0}\right) = s_0$$

$$\frac{St}{\sqrt{\pi}}\frac{e^{-s_0^2}}{\operatorname{erf}s_0} = s_0, \; St = \frac{c_i|T_{i0}|}{L} \tag{7.20}$$

where we have used the fact that T_{i0} must be less than 0 for the ice to freeze. In this case, we take $T_{i0} = -20$ °C. Then solving (9) numerically using the values in Appendix A, we have that:

$$s_0 = 0.237 \qquad (7.21)$$

To traverse the wick, the freezing front must advance a distance $R_m - R_v$, where R_m is the outer radius of the wick (the subscript "m" refers to "middle"), while R_v is the outer radius of the core (the subscript "v" refers to "vacuum"). See Figure 7.13 Hence the freezing front reaches the inner edge of the wick when:

$$(R_m - R_v) = 2s_0\sqrt{D_i t}$$

$$t = \frac{(R_m - R_v)^2}{4s_0^2 D_i} \qquad (7.22)$$

Then using the parameters in Appendix A, we have:

$$t = \frac{(2\,cm - 1.25\,cm)^2}{4(0.237)\left(1.23\times10^{-6}\,\frac{m^2}{s}\right)} = \frac{(7.5\times10^{-3}\,m)^2}{4(0.237)(1.23\times10^{-6}\,m^2)}$$

$$= \frac{(7.5)^2}{4(0.237)^2(1.23)}s = 204s \qquad (7.23)$$

TESTING

A. ACCEPTANCE TEST PROCEDURE

All heat pipes were subjected to acceptance testing as required per the JPL *Heat Pipe Specification* document. Acceptance testing consists of freeze/thaw testing and heat transport/thermal conductance characterization testing, performed in the order as listed. The procedures for the individual tests are outlined as follows.

B. FREEZE/THAW TESTING

Freeze/thaw testing was performed inside of a thermal chamber. All of the heat pipes were oriented vertically with the pinch-off end (evaporator) pointed downwards. The chamber was set to cycle from below −35 °C to above +85 °C for 100 cycles. The chamber controller was monitoring temperature of one heat pipe to ensure the low temperature and high temperature extremes were met during each cycle. Prior to freeze/thaw testing, the heat pipe diameters at the pinch-off end were measured and recorded. After completion of freeze/thaw testing, the heat pipe diameters at the pinch-off end were measured and recorded once again. All pipes were visually

FIGURE 7.20 Heat transport test fixture (front).

FIGURE 7.21 Heat transport test fixture (back).

inspected for any signs of deformation, in addition to verifying that the diameter of the heat pipe did not exceed 102% of the original dimension after freeze/thaw testing.

C. Heat Transport/Thermal Conductance Characterization Testing

Heat transport/thermal conductance characterization testing was performed on the benchtop in lab-ambient air pressure and temperature. All of the heat pipes were tested in an adverse gravitational orientation (evaporator above condenser).

The heat transport test fixture is shown in Figure 7.20 and Figure 7.21. The test fixture consists of evaporator heater blocks and condenser chiller blocks. The evaporator chiller blocks had two chip resistors that were used to apply the heat load to the backside of the block via DC power supply. The condenser chiller blocks had two flow-through paths for coolant at a set temperature to flow through. The test fixture was instrumented with plunger thermocouples, one in the center of the condenser chiller block (TCOND), and three along the length of the evaporator heat block (TEVAP1 at the top, TEVAP2 in the middle, and TEVAP3 at the bottom), as shown in Figure 7.22.

Parasitic heat loss was measured for the test fixture via installing a heat pipe in the fixture without any coolant flowing to the chiller block. The test fixture was insulated to prevent excessive heat loss. Power was then applied to the heat pipe until the pipe stabilized at the operational temperature for the test. The power applied (measured PHL) was then added to the nominal power level for the tests. This additional power was not included in the calculation of the thermal conductance of the heat pipe.

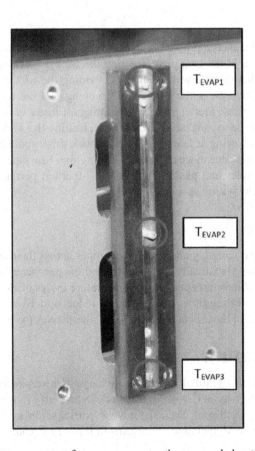

FIGURE 7.22 Heat transport test fixture—evaporator thermocouple locations.

D. POST-QUALIFICATION HEAT TRANSPORT/THERMAL CONDUCTANCE CHARACTERIZATION TESTING

Post-qualification heat transport/thermal conductance characterization testing was a repetition of the heat transport/thermal conductance characterization testing performed during acceptance testing. This was performed in order to verify heat transport capability post-qualification testing. The procedure for post-qualification heat transport/thermal conductance characterization testing is identical to the heat transport/thermal conductance characterization testing procedure listed previously.

FREEZE/THAW TEST RESULTS

Freeze/thaw testing was performed on all heat pipes per the acceptance test. The outer diameter of the heat pipe near the pinch-off end did not exceed 102% of the original dimension of the heat pipe due to freeze/thaw action.

CONCLUSION

A simple volume-balance approach was used to understand rupture due to ice expansion inside the wick of the heat pipe. Treating the wick as packed spheres, volume analysis showed that rupture was not likely providing the wick was loaded to less than 97% of saturation. Next, the problem of a moving ice front was tracked by solving a one- and two-phase Stefan problem, obtaining estimates of 3–5 minutes for the time to it takes the ice to traverse the wick. Thus, loading the wick to less than 97% of saturation and allowing at least 5 minutes of soak time would reduce the risk of rupturing the heat pipe due to ice expansion. Twenty-four heat pipes were put through 100 freeze/thaw cycles and passed pre- and post-thermal performance as well as dimensional stability within specified guidelines.

FUTURE WORK

Future work involves creating and testing heat pipes across thousands of freeze and thaw cycles to show statistically that the sintered copper water heat pipe will not fail. Current freeze/thaw tests performed temperature cycles of close to 1 hour; with these new insights, testing times can be greatly reduced to avoid costly testing that discourages the use of these heat pipes as viable alternatives for space applications.

ACKNOWLEDGEMENT

We like to extend our gratitude to Subas Acharya (University of Texas, Dallas), Valeria Barra (University of Colorado, Boulder), Dean Duffy (NASA), and Vrushaly Singlot (University of Texas, Dallas) for your participation in the Thirty-Fourth Annual Workshop on Mathematical Problems in Industry in Claremont, CA.

REFERENCES

1. Cheung, Kwok, "Flight qualification of copper water heat pipes at naval research laboratory," *38th AIAA Thermophysics Conference. American Institute of Aeronautics and Astronautics*, 2005.
2. Capri, "Sintered wick structured heat pipes," http://capri.co.in/sintered-wick-structured-heat-pipes/ [cited June 26, 2018].
3. Alexiades, V., and A.D. Solomon, *Mathematical Modeling of Melting and Freezing Processes,* Taylor & Francis, Boca Raton, FL, 1992.
4. Cverna, F., and ASM International. Materials Properties Database Committee, "ASM ready reference: Thermal properties of metals," Materials data series, ASM International, Materials Park, OH, 2002.
5. Engineering Toolbox, "Ice—thermal properties," www.engineeringtoolbox.com/ice-thermal-properties-d_576.html [cited June 26, 2018].
6. Engineering Toolbox, "Water—density, specific weight and thermal expansion coefficient," www.engineeringtoolbox.com/water-density-specific-weight-d_595.html [cited June 26, 2018].
7. Engineering Toolbox, "Water— specific heat," www.engineeringtoolbox.com/specific-heat-capacity-water-d_660.html?vA=20&units=C# [cited June 28, 2018].
8. Engineering Toolbox, "Water—thermal conductivity," www.engineeringtoolbox.com/water-liquid-gas-thermal-conductivity-temperature-pressure-d_2012.html [cited June 28, 2018].
9. Karditsas, Panayiotis J., and Marc-Jean Baptiste, "Pure copper," http://www-ferp.ucsd.edu/LIB/PROPS/PANOS/cu.html [cited July 6, 2018].
10. "The problem of freezing of copper water heat pipes," Juan Cepeda-Rizo, Jet Propulsion Laboratory Problem Participants Subas Acharya, University of Texas, Dallas Valeria Barra, University of Colorado, Boulder Dean Duffy, NASA David A. Edwards, University of Delaware Vrushaly Singlot, University of Texas, Dallas and others . . . Thirty-Fourth Annual Workshop on Mathematical Problems in Industry June 25–29, 2018 Claremont Graduate University

PHASE CHANGE γ-ALUMINA AQUEOUS-BASED NANOFLUID FOR IMPROVING HEAT PIPE TRANSIENT EFFICIENCY (THE NANO HEAT PIPE)

INTRODUCTION

Nanotechnology is making its way through the world as the next frontier in science. Many applications have been found and many more likely will make their way in the near future. Some people are expecting nanotechnology to be larger than the internet boom of the late 1990s [1]. Reducing the magnitudes of scale is enabling very large increases in performance that are otherwise difficult to achieve. Nanoparticles dispersed in a fluid created a new classification of fluids called nanofluids, a term made popular by Choi [2], who has studied extensively their properties, such as bulk thermal conductivity of fluids. Wen and Ding [3] have studied pool-boiling properties of gamma-alumina nanoparticles dispersed in an aqueous nanofluid and have conclusive data pointing to increases in nucleate boiling heat transfer of 40%. Many mathematical models are being investigated that quantify the attributes that the nanofluids possess, which cannot be explained by previous models based on meso-scale-size particles. Bhattacharya [4] and Jang [5] have proposed Brownian motion models to explain anomalous increases in thermal conductivity attributed to these nanoparticles. Recent work by Shi et al. [6] has created mathematical models to predict enhancements to pool boiling.

The subject of this work relates to use of a γ-alumina aqueous-based ternary nanofluid in a wickless heat pipe for increasing thermal performance. Pioneering work by Chien and Tsai [7] has shown appreciable increase in thermal performance of a disk-shaped miniature heat pipe (DMHP) used as a heat spreader for a laser diode transistor outline (TO) can package. As demands for low-cost, high-performance cooling for the electronics industry increases, a rectangular, wickless heat pipe, otherwise called a thermosyphon, is investigated.

EXPERIMENTAL STUDY

The rectangular thermosyphon is shown in Figure 7.23 and consists of a hollow rectangular brass tube filled with γ-alumina aqueous-based working nanofluid with an average particle size of 34.2 nm. The nanofluid was obtained from Nanophase, Inc. Three different mixtures were created using a 23% suspension created by Nanophase: they were 1%, 2%, and 5% by weight mixtures, which were prepared by diluting the suspension with deionized water and mixing in an ultrasonic bath for 15 minutes. The filling setup is shown in Figure 7.24 and consists of a premeasured syringe containing the nanofluid, a vacuum pump set consisting of both roughing and diffusion pumps, a cooling chuck, and a ultrasonic welder for hermetically sealing after filling. Figure 7.25 shows an actual picture of the finished rectangular thermosyphon.

The test setup is shown in Figure 7.26 and consists of a cylindrical vessel with a heater element and temperature control for maintaining the temperature of the water inside to a constant preset. The thermosyphon evaporator end is inserted vertically and submersed approximately 25 mm into the heated water. The outside ambient temperature is monitored, and the condenser end of the specimen is allowed to cool by natural convection. A thermocouple 10 mm from each crimped end monitors the temperatures at the condenser and evaporator sides of the thermosyphon. A data acquisition system monitors the temperature of the evaporator side versus time.

ANALYTICAL MODEL

Heat pipes in most applications tend to have low thermal capacitance and overall thermal resistance compared with the components that they are heating and cooling. In such applications, it may be acceptable to model the heat pipe as a transient lumped mass with a uniform temperature at an instant of time [8].

We start with heat equation with an added heat source:

$$\frac{\partial T}{\partial t} = \alpha \nabla^2 T + \frac{\dot{q}}{\rho c_p} \tag{7.24}$$

where T is the temperature of the thermosyphon, α is the thermal diffusivity equal to $\frac{k}{\rho c_p}$, where k is the thermal conductivity, ρ is the density, and c_p is the specific heat, and $\dot{q}$ is the heat source.

If we assume the thermosyphon to have a uniform temperature, a valid assumption for low levels of power, then the equation becomes:

$$\rho c_p \frac{\partial T}{\partial t} = \dot{q}_{sum} \tag{7.25}$$

where the $\dot{q}_{sum}$ is the sum of the heat sources entering and exiting the system.

Fahgri [9] presented a transient lumped heat pipe model that determines the average temperature as a function of time:

$$Q_e - (q_{conv} + q_{rad})S_C = C_t \frac{dT}{dt}, \tag{7.26}$$

where Q_e is the heat input, q_{conv} is the output heat flux by convection, q_{rad} is the radiation heat flux, S_C is the surface area around the circumference of the cooled section, C_t is the total thermal capacity of the system, and T is the assumed uniform temperature of the thermosyphon. The total thermal capacity is defined as:

$$C_t = \rho V_t c_p,$$

where ρ is the density, V_t is the total volume, and c_p is the specific heat of the system. For testing temperatures of the thermosyphon, the radiation heat flux can be neglected. Also, to simplify the formulation for a convective boundary condition, two external thermal resistances are defined as:

$$R_c = 1/(h_c S_c), \tag{7.27}$$

$$R_e = 1/(h_e S_e), \tag{7.28}$$

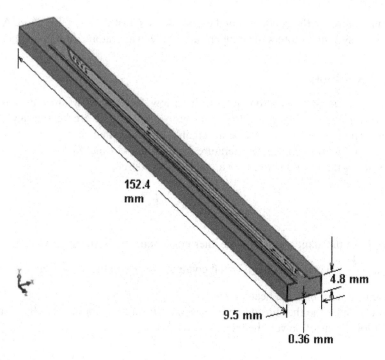

FIGURE 7.23 Rectangular thermosyphon filled with a γ-alumina aqueous-based working nanofluid.

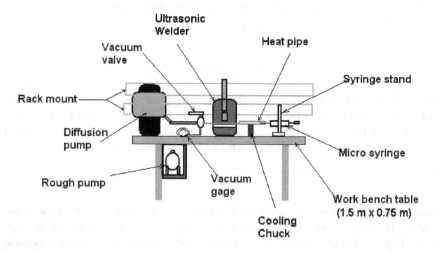

FIGURE 7.24 Thermosyphon charging station consisting of micro syringe, pumps, cooling chuck, and ultrasonic welder.

FIGURE 7.25 The rectangular thermosyphon after charging and sealing.

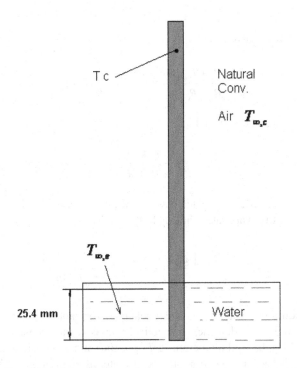

FIGURE 7.26 Experimental test setup consisting of heated water vessel and inserted vertical thermosyphon.

where h is the external heat transfer coefficient at the evaporator or the condenser and S is the surface area of the evaporator or condenser. Normally in a thermosyphon, R_e consists of resistance through the shell wall and liquid-to-vapor transition. In this analysis, we neglect the resistance through the thin shell of the thermosyphon.

The boundary conditions of the test could be described as follows:

$$T_{\infty,e} = T_{\infty,e1} \text{ for } t < 0$$

$$T_{\infty,e} = T_{\infty,e2} \text{ for } t \geq 0$$

$$T_{\infty,c} = T_{\infty,c1} = \text{constant}$$

where $T_{\infty,e1}$ is the initial evaporator ambient temperature.

The ambient temperature at the evaporator is a function of time, and the transient heat input is found using the thermal resistance between the evaporator and the environment. The initial condition is:

$$T(0) = \frac{T_{\infty,e1} R_c + T_{\infty,c1} R_e}{R_c + R_e} \tag{7.29}$$

where R_e is the evaporator thermal resistance given by (7.28). The energy equation is:

$$\frac{T_{\infty,e2} - T}{R_e} - \frac{T - T_{\infty,c1}}{R_c} = C_t \frac{dT}{dt} \tag{7.30}$$

The exact solution is given by

$$T(t) = \frac{T_{\infty,c1} R_e + T_{\infty,e1} R_c}{R_e + R_c} + \frac{R_c (T_{\infty,e2} - T_{\infty,e1})}{R_e + R_c} (1 - e^{-t/\tau}) \tag{7.31}$$

where the time constant is given by:

$$\tau = \frac{C_t R_c R_e}{R_c + R_e} \tag{7.32}$$

We can use the Rohsenow nucleate pool boiling equation to arrive at the convection coefficient at the evaporator end, h_e [10]:

$$q/A = \mu_L h_{fg} \left[\frac{g(\rho_L - \rho_V)}{\sigma} \right]^{1/2} \left[\frac{c_{pL}(T_S - T_{SAT})}{C_{sf} h_{fg} \, \mathrm{Pr}^{1.7}} \right]^3 \tag{7.33}$$

where μ_L is liquid viscosity, h_{fg} is latent heat of vaporization, g is gravity, ρ_L and ρ_V are liquid and vapor densities, respectively, σ is surface tension, c_{pL} is liquid specific heat, T_S is the wall temperature, T_{SAT} is the saturation temperature, and C_{sf} is the constant based on fluid type and wall material combination. The following parameters where taken:

$$\mu_L = 7.69E - 04(Pa \cdot s)$$
$$h_{fg} = 2.359E + 06(J/kg)$$
$$g = 9.81(m/s^2)$$
$$\rho_L = 983.3(kg/m^3)$$
$$\rho_V = 0.033(kg/m^3)$$
$$\sigma = 0.0709(N/m)$$
$$c_{pL} = 4,182(J/kgK)$$
$$T_S = 73\,^{\circ}C$$
$$T_{SAT} = 33\,^{\circ}C$$
$$C_{sf} = 0.013$$

$$T_{\infty,e1} = 30^{\circ}C, T_{\infty,e2} = 73^{\circ}C$$

$$T_{\infty,c1} = T_{\infty,c2} = 30^{\circ}C$$

Knowing that $h_e = \dfrac{q}{A(T_S - T_{SAT})}$,   (7.34)

we can plug (7.34) into (7.28) and then into (7.31) to have the complete transient response of the thermosyphon.

According to Grigull for free or natural convection, h_c is calculated by a classical approach using the correlation [11]:

$h_c = \dfrac{Nu \cdot k}{H}$ (7.35), H is the height of the plate and k is the thermal conductivity of air.

$$Nu = 0.55\sqrt[4]{Gr\,Pr}, \text{ for } 1700 < Gr\,Pr < 10^8 \qquad (7.36)$$

where Nu is the Nusselt number, Gr is the Grashof number. Also:

$$Gr = \frac{g\alpha\Delta\rho_L^{\,2}T \cdot H^3}{\mu_L^{\,2}}, \qquad (7.37)$$

where α is the thermal expansion coefficient and ΔT is the temperature difference.

Work by Fahgri has shown the lumped capacitance analysis is close to the more complicated nonlinear two-dimensional model under operating conditions similar to those in this experiment. As a result, the more complex numerical simulations to solve the 1D and 2D nonlinear equations were not pursued.

It is also important to note that the properties of specific heat, surface tension, and viscosity of the nanofluid are assumed to be close to those of pure water since the nanoparticles and the dispersant make up a small percentage of the total bulk volume.

EXPERIMENTAL RESULTS

Figure 7.27 shows the measured results of four thermosyphons, each with a different type of working fluid, and Figure 7.28 shows the predicted and measured values for the thermosyphon transient temperature response using Rohsenow nucleate boiling estimates and the lumped capacitance model. The measured results show a grouping pattern, where the 5% and 2% nanofluid thermosyphons have indistinguishable results, and the 1% and 0% exhibit similar behavior. This seems to point to instability or heightened sensitivity to fill ratios between 1% and 2%.

For the prediction of the nanofluid-filled thermosyphon, the thermal conductivity measurements by Das et al. [12] were used for an alumina/water nanofluid mixture

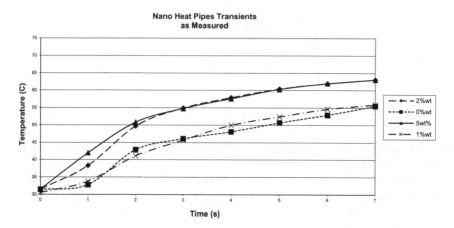

FIGURE 7.27 Measured transient thermal response, T_c, for 5%, 2%, 1%, and 0% nanoparticles by weight thermosyphons.

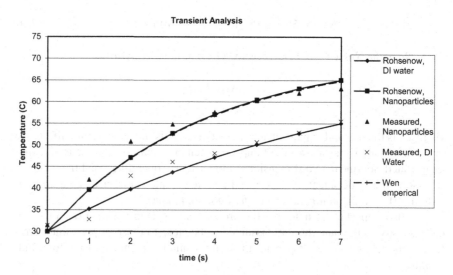

FIGURE 7.28 Predicted vs. measured temperature results, T_c, for both de-ionized water-filled and 2% by wt. nanofluid-filled thermosyphons.

at our operating temperatures. For 2% by weight of alumina nanofluid, these measurements showed a 15% increase in the thermal conductivity of the water [12]. This increased thermal conductivity was then inserted into the Rohsenow equation [10] by means of the Prandtl number. Wen and Ding also show enhanced nucleate boiling convection heat transfer with introduction of the nanoparticles of similar weight fraction. Their analysis showed 40% decrease in R_e, which we insert into the lumped capacitance model (7.30) for the nanoparticles thermosyphon transient estimates and show similar results to Rohsenow. It is apparent that the measured results showed a faster transient response than what was predicted. This could be due to the addition of a surfactant to improve dispersion in the thermosyphon versus the binary solution used by Wen and Yulong in their analysis. Many improved thermal results have been published that show enhanced particle dispersion improve the thermal properties of the nanofluid [12–16].

SUMMARY AND CONCLUSIONS

The results of the experiment are summarized as follows:

- Results showed the γ-alumina aqueous-based nanofluid improves the transient efficiency of the thermosyphon over 50% versus regular water.
- The ideal nanoparticles mix ratio for the nanofluid appeared to be 2% or slightly less based on measured results. There was a tremendous improvement in efficiency occurring between 1% and 2% by weight.
- Analytical results underestimated transient thermal response time by as much as 13% within the first 3 seconds of the test, but then correlated well after 6 seconds. Existence of a dispersant to improve particle dispersion may have attributed to this difference.

REFERENCES

1. Waters, R., "Why nanotechnology is the next big thing," *Financial Times, Ltd*, London, UK, March 29, 2005.
2. Choi, U.S., "Enhancing thermal conductivity of fluids with nanoparticles," in *Developments and Applications of Non-Newtonian Flows*, ASME, New York, 1995, pp. 99–105.
3. Wen, D., and Y. Ding, "Experimental investigation into the pool boiling heat transfer of aqueous based g-alumina nanofluids," *Journal of Nanoparticle Research*, vol. 7, 2005, pp. 265–274.
4. Bhattacharya, P., S.K. Saha, P.E. Phelan, and R.S. Prasher, "Brownian dynamics simulation to determine the effective thermal conductivity," *Journal of Applied Physics*, vol. 11, 2004, pp. 6492–6505.
5. Jang, S.P., and U.S. Choi, "Role of Brownian motion in the enhanced thermal conductivity of nanofluids," *Applied Physics Letters*, vol. 84, no. 21, 2004, pp. 4316–4318.
6. Shi, M., Y. Zhao, and Z. Liu, "Study on boiling heat transfer in liquid saturated particle bed and fluidized bed," *International Journal of Heat and Mass Transfer*, vol. 46, 2003, pp. 4695–4702.
7. Chien, Hsin-Tang, and Chien-In Tsai, "Improvement on thermal performance of a disk-shaped miniature heat pipe with nanofluid," *International Conference Electronic Packaging Proceedings—Institute of Electrical and Electronics Engineers, ICEPT IEEE*, 2003, pp. 389–391.
8. Colwell, G.T., and J.M. Modlin, "Mathematical heat pipe models," *8th International Heat Pipe Conference*, vol. 1, 1992, pp. 162–166.
9. Fahgri, A., *Heat Pipe Science and Technology*, Taylor and Francis, New York, 1995, pp. 267–274.
10. Welty, J.R., C.E. Wicks, and R.E. Wilson, *Fundamentals of Momentum, Heat, and Mass Transfer*, 3rd Ed., John Wiley & Son, Hoboken, NJ 1984, pp. 386–390.

11. Gieck, K., and R. Gieck, *Engineering Formulas*, 6th Ed., Gieck Publishing, West-Germany, 1990, 2011.
12. Das, S.K., and N. Putra, "Temperature dependence of thermal conductivity enhancement for nanofluids," *Journal of Heat Transfer-ASME*, vol. 125, 2003, pp. 567–574.
13. Yu, W., and U.S. Choi, "The role of interfacial layers in the enhanced thermal conductivity of nanofluids: A renovated Maxwell model," *Journal of Nanoparticle Research*, vol. 5, 2003, pp. 167–171.
14. Eastman, J.A., and U.S. Choi, "Enhanced thermal conductivity through the development of nanofluids," *Materials Research Society Symposia Proceedings*, vol. 457, 1997, pp. 3–12.
15. Eastman, J.A., and U.S. Choi, "Anomalously increased thermal conductivities of ethylene glycol-based nanofluids containing copper nanoparticles," *Applied Physics Letters*, vol. 78, no. 6, 2001, pp. 718–720.
16. Xuan, Y., Q. Li, and W. Hu, "Aggregation structure and thermal conductivity of nanofluids," *American Institute of Chemical Engineers Journals (AIChE)*, vol. 49, no. 4, 2003, pp. 1038–1043.

8 Convective Cooling of Semiconductors Using a Nanofluid

This chapter shows the principles of convection and how they are applied for pumped fluid loop cooling used for high-performance cooling of electronics. Pumped fluid loops approach temperature control in a different way and may not always be ideal for space due to the moving parts in the pump. They are valuable, however, during the ground testing campaigns, as pumped liquid nitrogen is used to cool the shrouds of an environmental chamber to simulate cold space temperatures.

INTRODUCTION

The concept of using direct liquid cooling over hot devices (e.g., immersion cooling) is not new; indeed, it has been around since dielectric fluids first became popular in the 1970s [1]. Much work was done in the late 80s by researchers such as Mudawar [2–4] on enhanced surfaces and forced convection. It was realized that limits of cooling performance were often hindered by pool boiling over the hot devices, which tended to insulate the device from the cooling fluid [5]. Tests have also shown that immersion cooling of bare die at typical flow rates around 0.3 gpm provides low performance (under 14 W/cm^2).

Two dies attach thermal interface materials, Reactive NanoTechnologies (RNT) solder foil and Loctite QMI 9501 silver-filled epoxy, were investigated for use in direct liquid immersion cooling of multichip modules (MCM): the goal of the immersion cooling feasibility experiment was to prove suitability for cooling devices in excess of 100 W/cm^2, while maintaining a junction temperature below 85 °C. The test vehicle used a test multichip module fabricated by International Business Machines (IBM), which consists of three resistance temperature device (RTD) thermal dies mounted on a low temperature co-fire ceramic (LTCC) substrate. The perceived advantages of an all-metal interface are two-fold:

- Improved mechanical attachment reliability (better material compatibility), and
- Improved thermal performance.

The improved thermal performance can again be bounded by first-order calculations but is driven largely by the bulk conductivity of the material. The epoxy has a published thermal conductivity of 7.3 W/mK and solidifies with a 76.2 micron (3 mil) thickness. With the given die size of 14.2 mm × 14.2 mm, this translates to an

DOI: 10.1201/9781003247005-8

effective resistance of 0.0518 °C/W or a 2.6 °C temperature difference with a 50 W heat flow. The metal interface has an overall thickness of about 160 microns (0.006 inch) and a thermal conductivity of about 50 W/mK, yielding an effective resistance of 0.0159 °C/W. With a 50 W heat flow this yields a 0.8 °C temperature difference.

FORCED CONVECTION FLOW

The relation of a fluid's properties to its efficiency as a coolant is linked to the convection heat transfer coefficient. The transfer of thermal energy by convection from a heated surface to a fluid in motion over the surface is determined by Newton's law of cooling, which may be expressed as:

$$q = \bar{h} A_s (T_s - T_f) \tag{8.1}$$

where $q, \bar{h}, A_s, T_s, T_f$ are cooling rate, average convection coefficient, surface area, surface temperature, and fluid temperature, respectively.

In electronic cooling, one goal is to maximize the ratio of heat rate to the temperature difference. That is, for an allowable temperature difference, it is desirable to maximize the amount of heat that can be dissipated. A large convection coefficient corresponds to a small convection resistance, and the two quantities are related by expressions of the form:

$$R_{th} = \frac{1}{\bar{h} A_s} \tag{8.2}$$

where R_{th} is the thermal contact resistance.

If cooling is by forced convection, a dimensionless form of the convection coefficient, termed the Nusselt number Nu, typically depends on a dimensionless flow parameter, termed the Reynolds number Re, and the Prandtl number Pr [6]. That is:

$$Nu_{Lo} \sim Re_{Lo}^m Pr^n \tag{8.3}$$

where m and n are positive exponents less than unity, $Nu_{Lo} \equiv h_{Lo}/k$, $Re_{Lo} \equiv Uo_{Lo}/v$, and Lo and Uo are the characteristic length and velocity, respectively, associated with the flow. The convection coefficient h is, therefore, governed by the following relations:

$$h \sim \frac{F_F U_o^m}{L_o^{1-m}} \tag{8.4}$$

where the *figure of merit* F_F, which combines the influence of all pertinent fluid properties, is:

$$F_F = \frac{k^{1-n} \rho^m c_p^n}{\mu^{m-n}} \tag{8.5}$$

where ρ, c, μ are density, specific heat, and dynamic viscosity, respectively.

The convection coefficient increases with F_F and, hence, with increasing thermal conductivity k and Prandtl number Pr, as well as with decreasing kinematic viscosity. Because $v \equiv \mu/\rho$ and $Pr \equiv v/\alpha = c_p\mu/k$, F_F may also be expressed as:

$$F_F = \frac{k^{1-n}\rho^m c_p^n}{\mu^{m-n}} \qquad (8.6)$$

Hence, the convection coefficient also increases with increasing density ρ and specific heat cp, as well as with decreasing dynamic viscosity μ.

Values of n are typically in the range $0.33 \leq n \leq 0.40$, and $m = 0.80$ for *turbulent flow*. For *laminar, external flow* over a flat surface, $m = 0.5$, whereas for the special case of *laminar, fully developed, internal flow* in a duct, $m = n = 0$ and $F_F = k$.

OVER PROTRUDING HEATED SURFACES

It was determined that flow in a closed channel over slight protrusions would be a good model. The effect of slight protrusions was considered experimentally for a single heat source, and the results were compared with those for a flush-mounted source [6]. When the comparison was based on the heat rate per exposed surface area, which included the sides of the protruding source, heat transfer enhancement resulting from the protrusion was less than 5%.

Experiments were conducted for FC-72 flowing in a vertical channel of width $W = 20$ mm, with four in-line simulated chips of length $L_h = 10$ mm and spacing $S_L = 10$ mm [7]. Protrusion and channel heights of $Bh = 0, 0.2L_h(0, 2$ mm), and $H = Lh(10$ mm) were considered for Reynolds numbers in the range $2.2 \times 10^2 < Re_{Lh} < 6.3 \times 10^4$. The Nusselt number was defined as $\overline{Nu}_L = QL_h / A_{ex}(T_h - T_{m,i})$,

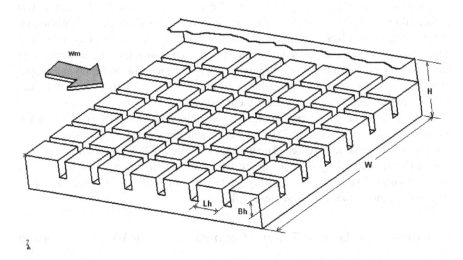

FIGURE 8.1 Channel flow over a heat sink directly attached to a chip modeled as an in-line array of protruding heat sources.

and the mean velocity was determined from the actual cross-sectional area of the flow, $w_m = \dot{m} / \rho(WH - B_h L_h)$. For the first chip ($j = 1$) and Bh = 0, the data were correlated by:

$$\overline{Nu_L} = C \operatorname{Re}_{L_h}^m \operatorname{Pr}^{0.33} \tag{8.7}$$

with C = 0.397 and m = 0.617. The correlation over-predicts the results of Maddox and Mudawar by approximately 10%. The Nusselt number decreased from the first to the second row, with a nearly fully developed condition being reached after the second row and $\overline{Nu_L}$ for the first row exceeding that of the last row by the first to the second row, and results for the second through the fourth row were correlated by (8.7), but with $C = 0.367$ and $m = 0.609$. The correlation is in good agreement with that of Gersey and Mudawar. For the downstream rows, there was little difference between results for $B_h = 0$ and $B_h = 2$ mm. For the first row, results for $B_h = 2$ mm exceeded those for $B_h = 0$ by less than 20% for $Re_{Lh} < 10^4$.

NANOFLUIDS FOR COOLING ELECTRONICS

The conventional approach for increasing cooling rates is the use of extended surfaces such as fins and microchannels. However, current designs have already stretched this approach to its limits. Therefore, there is an urgent need for new and innovative concepts to achieve ultra-high-performance cooling [8]. Equation (8.6) indicates that the figure of merit F_F is directly proportional to the convection coefficient, which is proportional to the conductivity of the fluid. A gain in the cooling performance, therefore, could be directly accomplished by increasing the thermal conductivity of the fluid. To achieve this, a new class of heat transfer fluids is being developed by suspending nanocrystalline particles in a liquid such as water or oil. The resulting "nanofluids" possess extremely high thermal conductivities compared to the liquids without dispersed nanocrystalline particles [9]. Much work has been published on water and ethylene glycol based nanofluids, but very little on fluorocarbon dielectric fluids, probably due to dispersion difficulties of the nanoparticles. The preparation of the nanofluid must ensure proper dispersion of nanoparticles in the liquid and proper mechanism such as control of pH value or addition of surface activators to attain the stability of the suspension against sedimentation [10].

In this experiment, a ternary nanofluid consisting of 34.2 nm Al_2O_3 nanoparticles, HFE-7100 dielectric fluid from 3M, and laurate salts emulsion to enhance dispersion was created in house and studied. The mixture consisted of 0.3% by volume nanoparticles and less than 1% by volume laurate salts emulsion mixed in an ultrasonic bath for 15 minutes. The result was a milky-white nanofluid with good apparent dispersion of the nanoparticles in the dielectric fluid.

ESTIMATION OF THE EFFECTIVE THERMAL CONDUCTIVITY OF THE TERNARY NANOFLUID

The first formal study of thermal conductivity enhancements by introducing suspended particles can be attributed to Maxwell [11] over 100 years ago. Maxwell's model predicts that the effective thermal conductivity of suspensions containing spherical particles increases with the volume fraction of the solid particles [12]. Work by Hamilton and Crosser [13] looked at the benefit of increased particle surface area for non-spherical particles and modified Maxwell's model accordingly. Both the Maxwell model and modified Maxwell models by Hamilton and Crosser did well in estimating effective thermal conductivities of meso-scale and micro-scale particles slurries, within one to two orders of magnitude of estimating measured values of nanofluids. This discrepancy is partially attributed to the lack of representation of a large increase of the surface area to volume that nanoparticles possess. For instance, the surface area to volume is 1000 times greater for a 1 nm particle than a 1μm particle.

Maxwell model [11]:

$$k_{Maxwell} = \frac{k_p + 2k_l + 2(k_p - k_l)\phi}{k_p + 2k_l - (k_p - k_l)\phi} k_l \tag{8.8}$$

where k_p and k_l are the thermal conductivity values of the particle and liquid, respectively, and ϕ is the volume fraction of particles in the fluid. The following values are typical:

k_p = 30 W/m²K
k_l = .115 W/m²K
ϕ = .003

Modified Maxwell Model[12]:

$$k_e = \frac{k_{pe} + 2k_l + 2(k_{pe} - k_1)(1+\beta)^3 \phi}{k_{pe} + 2k_l - (k_{pe} - k_1)(1+\beta)^3 \phi} k_l \tag{8.9}$$

and

$$k_{pe} = \frac{\left[2(1-\gamma) + (1+\beta)^3(1+2\gamma)\right]\gamma}{-(1-\gamma) + (1+\beta)^3(1+2\gamma)} k_p \tag{8.10}$$

where k_{pe} is the equivalent thermal conductivity of the equivalent particle, and where:

$$\gamma = k_{layer}/k_p = 10 k_1/k_p$$

$$\beta = \delta_T/r,$$

where δ_T = layer thickness = 10 nm, r = particle radius, = 17.1 nm.

The parameter k_{layer} is the estimated thermal conductivity of the Helmholtz-like layer around the nanoparticles and is estimated to be 10 times the thermal conductivity of the base liquid. This layer acts as a quasi-solid due to its ordered structure and is believed to have a much higher conductivity than the bulk fluid. The value k_{pe} is the effective thermal conductivity of the particle surrounded by a Helmholtz-like layer of liquids that surrounds each particle. k_{pe} is dependent on the size of the layer and the ratio of layer thickness to particle size, or $\beta = h/r$, layer thickness, h, and particle radius, r. The modified equation gives good predictions for nanoparticles 10 nm or less with a layer thickness 1–2 nm, but adds little distinction from the original Maxwell model for particles in the range of 30 nm or greater.

BROWNIAN MOTION MODEL

Jang and Choi [14] offer an alternative estimate of the thermal conductivity by investigating the kinetic motion of particles and liquid molecules, Brownian motion, and extracting a value derived by the Kapitza resistance. The motion model is divided into four modes of interaction: (1) collisions between fluid molecules, (2) thermal diffusion of nanoparticles in fluids, (3) collisions between nanoparticles, and (4) thermal interaction of dancing nanoparticles with base fluid molecules. The fourth mode is defined by:

$$J_U = h(T_{nano} - T_{BF})f = h\delta_T f \frac{(T_{nano} - T_{BF})}{\delta_T} \sim -h\delta_T f \frac{dT}{dz} \qquad (8.11)$$

where h is the heat transfer coefficient for flow past nanoparticles. Neglecting the third mode, the following can be derived:

$$k_{eff} = k_{BF}(1 - f) + k_{nano}f + fh\delta_T \qquad (8.12)$$

Where k_{eff} is the effective thermal conductivity of the particle/liquid system, k_{BF} is the bulk conductivity of the fluid, and f is the volumetric ratio particles to liquid. For flow past a sphere, the Nusselt number is:

$$Nu = 2.0 + 0.5RePr + O(Re2Pr2). \qquad (8.13)$$

For typical nanofluids, the Reynolds and Prandtl numbers are on the order of 1 and 10, respectively, and (8.6) can be simplified as:

$$Nu \sim Re^2 Pr^2 \qquad (8.14)$$

The heat transfer coefficient for flow past nanoparticles is then defined by:

$$h \sim \frac{k_{BF}}{d_{nano}} Re_{d_{nano}}^2 Pr^2 \qquad (8.15)$$

$$\text{where } Re_{d_{nano}} = \frac{\overline{C_{R.M.}} d_{nano}}{v}, \qquad (8.16)$$

where v is the kinematic viscosity and $\overline{C_{R.M.}}$ is the random motion velocity:

$$\overline{C_{R.M.}} = \frac{D_o}{\ell_{BF}}$$
(8.17)

where the nanoparticles diffusion coefficient is given by Einstein as:

$$D_o = \frac{k_b T}{3\pi\mu d_{nano}} \text{ and } k_b = 1.3807 \times 10^{-23} J / K \text{ is the Boltzmann constant. (8.18)}$$

and ℓ_{BF} = 1.4 nm is the mean free path distance, estimated as the fluid molecular diameter.

THERMAL CONDUCTIVITY MEASUREMENT

Modified hot wire transient techniques [15] can nondestructively and accurately measure the thermal conductivity and thermal effusivity of a material in seconds. The modification compared to the hotwire technique is that the heating elements are supported on a backing, which provides mechanical support, electrical insulation,

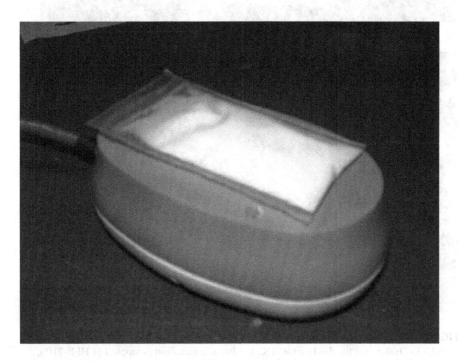

FIGURE 8.2 The nanofluid sits on top of the hot wire device for thermal conductivity measurement.

and thermal insulation. Such modification eliminates the intrusive nature of the hot wire technique, thereby allowing solids to be tested without melting or otherwise modifying the sample to conform to the geometry of the test cell. In the modified hot wire transient technique, one side of the heating element is in contact with the sample, while the other side is in contact with the backing material. During the testing, a constant current or voltage electrical source supplies power to the heating element to generate a one-dimensional heat flow, which is perpendicular to the testing surface of the sample. At the same time, the middle heating element also serves as the sensor by functioning as a resistance thermocouple.

The rate of the temperature rise at the sample-sensor interface can therefore be monitored in real time by recording the rate of the voltage change in a constant current configuration or recording the rate of the current change in a constant voltage configuration. Thermal conductivity and thermal effusivity are inversely proportional to the rate of temperature rise at the sample/sensor interface, because the speed of the heat conducted away from the interface is proportional to their values. Both thermal conductivity and thermal effusivity of the sample can thus be determined by correlating the interfacial temperature rise over time to a set of calibration standards tested under identical conditions. The modified hot wire transient technique therefore is a comparative (secondary) method of measurement, and the sensor should

FIGURE 8.3 Fluid samples for thermal conductivity testing. The fluid on the left (clear) is pure HFE-7100, while the fluid on the right (white) is a nanofluid composed of HFE-7100, 0.3% by volume aluminum oxide nanoparticles, and a small amount of laurate salts to aid in dispersion.

TABLE 8.1

Results of the Hot Wire Thermal Conductivity Measurements

Sample	Measured Thermal Conductivity (W/mK))	Mean Thermal Conductivity (W/m•K)
White Liquid	0.140	0.141
White Liquid	0.141	
White Liquid	0.141	
Clear Liquid	0.115	0.115
Clear Liquid	0.115	
Clear Liquid	0.115	

TABLE 8.2

Predicted vs. Measured Thermal Conductivity Values

	Thermal Conductivity	Normalized Conductivity	Error (predicted vs. measured)
Maxwell	0.116	1.009	17.71%
Modified Maxwell	0.119	1.036	15.55%
Brownian	0.142	1.235	0.89%
Measured Value	0.141	1.226	

be calibrated with known materials to perform thermal conductivity/effusivity measurements. Also, the modified hot wire transient technique treats the sample like a semi-infinite medium. As a result, it is required that the heat does not completely pass through the sample during the test.

Table 8.1 shows the measured thermal conductivity of both plain fluid and nanofluid using the hot wire method. The Maxwell model (8.8) resulted in a value of 0.116 W/mK, or a normalized conductivity value $k_{eff}/k_1 = 1.0086$, where k_{eff} is the effective thermal conductivity of the ternary system and k_1 is the thermal conductivity of the bulk liquid. The Choi et al. modified Maxwell model (8.10) resulted in a normalized conductivity value of 1.036, which is larger than the Maxwell model, but fails to correlate well with the measured value of 1.226. The Brownian motion model (8.12) had a normalized conductivity value of 1.235, which is comes very close to the measured value, or within 0.89% error.

Maxwell and modified Maxwell models are conduction based and known to have weakness in predicting thermal conductivity enhancements at the nanoscale [16]. Brownian motion takes into effect the random movements and collisions of small particles with each other and with the fluid molecules at the nanoscale, usually neglected by the conduction-based models. This is maybe why the latter better predicts the effective thermal conductivity of nanofluids.

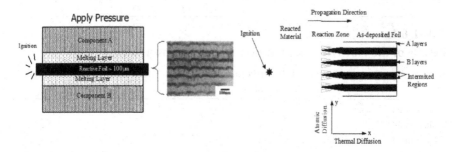

FIGURE 8.4 Reactive foil and bonding process.

NANOSCALE THERMAL INTERFACE

The challenges today with implementing a metallic bond relate to cost, the required processing temperatures, and the fact that reflow ovens are not well suited for the large heat sink mass of fully populated boards. Alternative methods to solder to silicon have been studied but are not well suited for heat sink or lid attachment. The reactive foil used in this work belongs to a new class of nano-engineered materials, in which controlled self-propagating exothermic reactions can be initiated and enable the room temperature soldering of components. The foil itself is a multilayer construction that is inserted between two solder layers and two components. A micrograph of the foil and a pictorial description of this bonding process are shown in Figure 8.4.

The reactive foil essentially provides a local heat source that can be substituted for a reflow-type soldering system. The bonding process is fast (< 1 second) but does require proper component preparation (metallization or prewet) and an ignition source. The surface preparation is dependent on several design variables such as the type of material (e.g., type of metal or ceramic) and the solder preference. Figure 8.5 illustrates some examples of lead-free options specific to attaching aluminum to various chip and package materials.

The attachment scenarios shown, as well as the application work reported in this chapter, focus on attaching a heat sink to an integrated circuit (IC) die or package. However, the ability to achieve a metallic bond between two components of different materials has application outside of this work, such as in package lid attach.

TEST VEHICLE DESIGN

The test vehicle was designed in accordance to a previous adopted test vehicle that was used to test an indirect liquid cooling module designed for cases where a cold plate was attached to the substrate of the device as a substitute for the heat spreader lid. The concern with immersion cooling was to make sure the liquid was properly routed to the chips, since elimination of the heat spreader would prevent any averaging of the heat before entering the liquid, which is the case for cold plates. Simulations were performed using a computational fluid dynamics (CFD) program called Floworks by SRAC to aid in designing a baffle chamber that best distributed the fluid over the three hot devices.

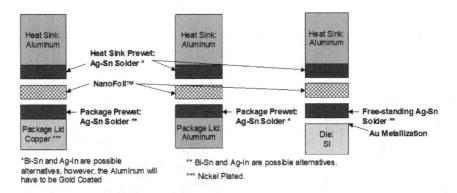

*Bi-Sn and Ag-In are possible
alternatives, however, the Aluminum will
have to be Gold Coated

** Bi-Sn and Ag-In are possible alternatives.

*** Nickel Plated.

FIGURE 8.5 Lead-free attachment methods for aluminum.

The first version, shown in Figure 8.6, used a serial distribution where the fluid cooled each device in series. The expected advantage was a means to provide the most coolant over each device, thus the potentially highest amount of heat flux removal. The second version baffle, shown in Figure 8.7, utilized parallel flow over the devices to best achieve uniform temperature distribution over the devices. The simulations showed that the parallel scheme provides similar to better overall heat flux removal than does the series scheme, with a much better temperature distribution across all the devices. The parallel scheme was selected for testing.

EXPERIMENTAL SETUP

An MCM thermal test vehicle was used in this study that is comprised of three RTD thermal dies mounted on an LTCC ceramic substrate. For the reactive-soldering process, the heat sink and dies were plated with Ti-Ni-Au, which allows for wet-ability of the surfaces. 50-micron Pb-Sn solder sheets were then placed on each surface with a 60-micron thick NanoFoil™ sandwiched in between. The foil was ignited with this assembly under an applied pressure of 50 N (11 lb).

The heat sinks were machined out of aluminum nitrate (AlN) and mounted on the 14.2 mm dies with the thermal interface material. The AlN material was chosen because of its thermal properties and because its thermal expansion matched well with silicon. The MCM with the attached heat sinks was then attached to the test or fan-out printed circuit board (PCB) with a transparent polycarbonate lid and bolster plate as shown in Figure 8.8.

The immersion cooling lid and bolster plate shown in Figure 8.8 use six shoulder screw-and-spring assemblies to compress the MCM against the test (fan-out) PCB. A compressible interposer-type connector identified in diagram in Figure 8.8 electrically connects the MCM substrate to the PCB. The remaining setup is shown in Figure 8.9. The signals and power heaters on the die are routed through the substrate and PCB, brought off the PCB through an edge connector, and finally connected to a switch box. The switch box interfaces to power supplies and temperature indicators and allows the heater power to be varied. A chiller with adjustable flow rate and

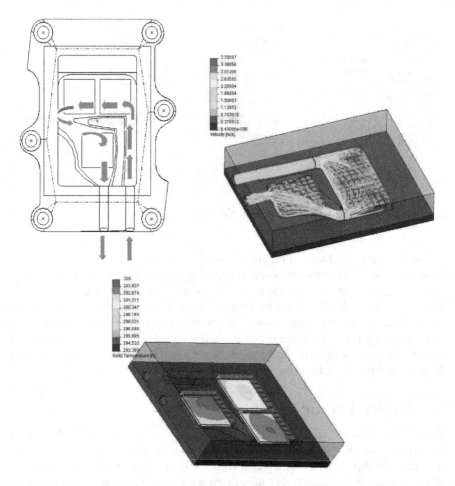

FIGURE 8.6 Layout and simulation results of the serial flow baffle. Notice the serial nature of the flow and how the first chip receiving fluid is much cooler than the last.

coolant temperature settings is used to pump the dielectric fluid (HFE-7100 from 3M) through the cooling lid and over the dies with the attached heat sinks.

EXPERIMENTAL RESULTS

Experiments were run with the setup shown in Figures 8.8 and 8.9 to assess the impact the heat sink interface material has on the die temperature. The results of the first tests are shown in Figures 8.10 through 8.12 and illustrate the die temperature sensitivity to power dissipation for the epoxy and all-metal interface. The tests were performed at three different flow rates. At the medium flow rate of 0.35 gpm, there is a noticeable reduction in the A3 junction temperature with the all-metal bond at power levels greater than approximately 40 watts, and after about 60 watts for the A2 die. This is believed attributed to the earlier inception of boiling that occurs with

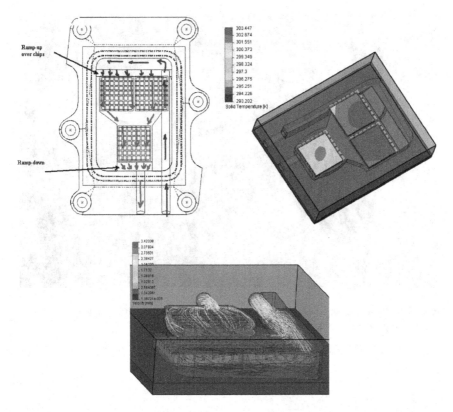

FIGURE 8.7 Layout and simulation results for the parallel flow baffle. The uniform flow over the top row of chips creates a better overall temperature distribution than does the serial flow scheme.

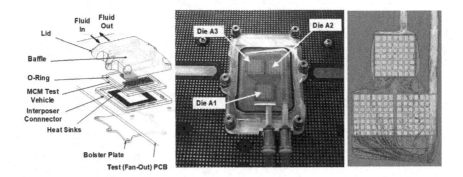

FIGURE 8.8 Direct immersion cooling test setup consisting of liquid-sealing lid, test MCM, connector, and printed circuit board. Flow pattern depicted at the far right for 0.35 gpm.

FIGURE 8.9 Immersion cooling test bench setup showing fan-out PCB (center) attached to panel box (upper right corner) and chiller unit (lower left corner).

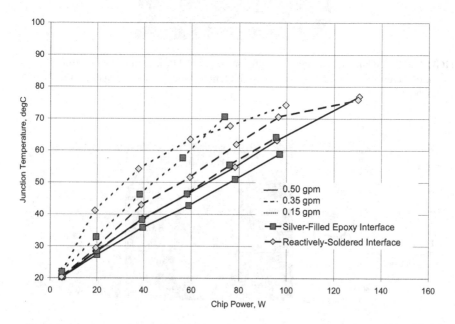

FIGURE 8.10 Average junction temperature of die A1 versus chip power level for both all-metal and epoxy thermal interfaces. The power level indicated is per die.

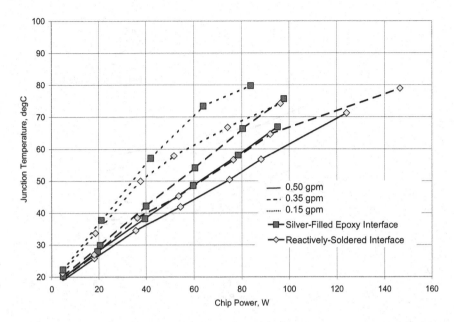

FIGURE 8.11 Average junction temperature of die A2 temperature versus chip power level for both all-metal and epoxy thermal interfaces. The power level indicated is per die.

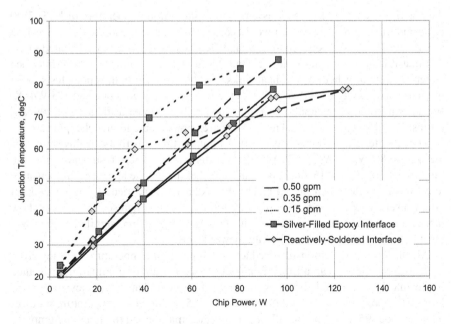

FIGURE 8.12 Average junction temperature of die A3 temperature versus chip power level for both all-metal and epoxy thermal interfaces. The power level indicated is per die.

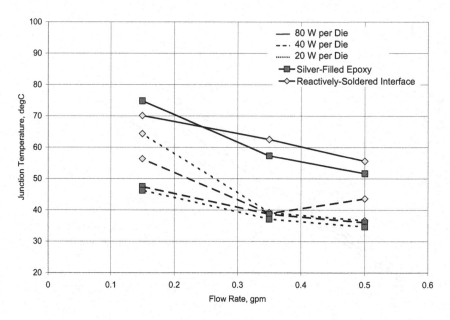

FIGURE 8.13 Average junction temperature of die A1 versus coolant flow rate for both all-metal and epoxy thermal interfaces.

the reactively soldered heat sink assembly (the saturation temperature of HFE-7100 is around 61 °C to 65 °C), which would cause a marked reduction in the convection resistance. The A1 die behaves somewhat differently, which is likely attributed to the fact that it is downstream of dies A2 and A3. It is worth noting that both interface materials were attached to AlN heat sinks, but the one used in the epoxy tests had a thermal conductivity of 190 W/mK, and the reactively soldered one had 175 W/mK. In addition, some of the nonlinearity may be attributed to parasitic losses to the air, for instance through the PCB. These losses are cumbersome to model explicitly.

The next testing assesses the junction temperature sensitivity to flow rate normalized to 80 W, 40 W, and 20 W of power dissipation. These results are shown in Figures 8.13 through 8.15. One observation is that the A1 die appears to be sensitive to the upstream flow patterns defined by dies A2 and A3, particularly at the medium power levels. Dies A2 and A3, which are both at the inlet and see fresh fluid, show more consistent temperature trends with flow rate and a more significant improvement by the metal interface at the high power levels.

All of these tests demonstrate reductions in junction temperature in going from a silver-filled epoxy to an all-metal one when a significant or dominating resistance is the interface material. From Figure 8.15 it is clear that at the elevated power levels, the epoxy interface module exceeds the 85 °C junction temperature requirement at about 95 W, while the all-metal interface maintained the junction temperatures below this maximum temperature limit even at 125 W. This improved cooling capacity allows for greater power dissipation potential in the existing and future designs.

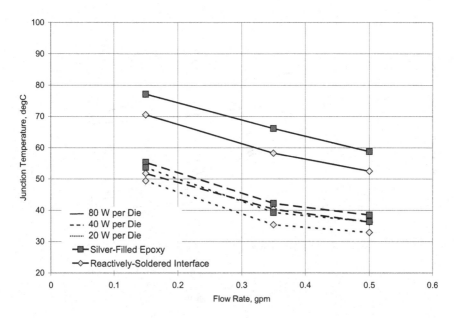

FIGURE 8.14 Average junction temperature of die A2 versus coolant flow rate for both all-metal and epoxy thermal interfaces.

PREDICTED VALUES, SINGLE-PHASE CONVECTION COOLING

For fluid values associated with HFE-7100 dielectric cooling flood (μ = 1.4E-03 kg/ms, ρ = 1500 kg/m^3, c_p = 100 J/kgK) and a 0.35 gpm flow rate), the resultant Reynolds and Prandtl numbers were 47,325 and 24.1, respectively. Inserting these values into (8.7), with m = .609 and C = .367 as suggested by Gersey and Mudawar, a predicted convective coefficient was found, h = 15,897 W/m^{2}K. With a heat rate value of 40 W, and using material conductivity values for the nanoscale thermal interface and AlN heat sink of 70 and 175 W/mK, respectively, resistance values can be calculated for heat conduction through heat sink and die with use of (8.2). The estimated junction temperature for die A2, T_s, could therefore be calculated by solving (8.1), which resulted in T_s = 33.7 °C. Figure 8.14 shows an experimentally arrived value of 32.9 °C for 0.35 gpm and all-metal nanoscale thermal interface case, so the resulting predicted versus measured error is 2.4% as shown in Table 8.3. This correlation between predicted and measured values fared well up to ~ 60 W of heat load, where second-phase change cooling associated with boiling of the liquid becomes pronounced and could not be neglected.

RESULTS OF PLAIN FLUID VS. NANOFLUID

In the last leg of the experiment, the test vehicle with the nanoscale all-metal thermal interface was rerun at a flow rate of 0.35 gpm with the dielectric ternary nanofluids and the results are shown in Figure 8.16. The average junction temperature with the

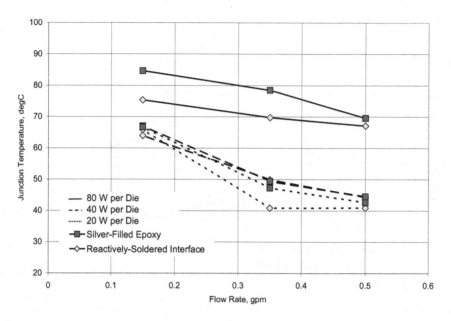

FIGURE 8.15 Average junction temperature of die A3 versus coolant flow rate for both all-metal and epoxy thermal interfaces.

TABLE 8.3
Junction Temperature of Die A2, Prediction vs. Measured

	Prediction	Measured	Error (predicted vs. measured)
Junction Temperature (Ts)	33.7 °C	32.9 °C	2.4%

nanofluids of die A2 had a maximum of 65.75 °C at 127 watts of power, versus the previously reported 79.41 °C temperature for the plain fluid case, or a 17.21% gain in performance.

The improvement in performance with the nanofluids was apparently similar to that seen by increasing the flow rate from 0.35 gpm to 0.5 gpm, though slightly better.

SUMMARY AND CONCLUSIONS

The results of the experiment are summarized as follows:

- The immersion cooling experiment with standard dielectric cooling fluid and conductive epoxy thermal interface resulted in 115 W at 85 °C and 0.35 gpm, equaling a heat flux of 55 W/cm^2.

- With the nanoscale thermal interface, 147 W at a junction temperature of 79.2 °C was achieved at die A2 with a flow rate of .35 gpm, translates into 70 W/cm². Current limitations of the MCM interposer prevented testing at a higher power. Projections indicate that 210 W should be easily attainable at a junction temperature under 85 °C, equaling a heat flux of near 99.8 W/cm² at the same flow rate.
- Addition of the ternary nanofluid improved performance by another 17.2%, thus increasing the heat flux to 117 W/cm² for the nanoscale thermal interface-ternary nanofluid system. The total performance improvement over the epoxy thermal interface and standard dielectric fluid was 113%.

This work demonstrates the implementation and thermal performance of an all-metal nanoscale interface achieved with a novel bonding process based on a reactive foil. The all-metal interface bond by its very nature offers lower thermal impedance than epoxy bonds do and solder-like bond strength. Introduction of the ternary dielectric nanofluids further reduced the device junction temperature to an appreciable amount. These attributes are valuable to the package-level heat sink application and die-level immersion cooling application presented in this work, and they offer similar benefits to other applications, particularly compact electronic systems such as blade servers and mobile computers.

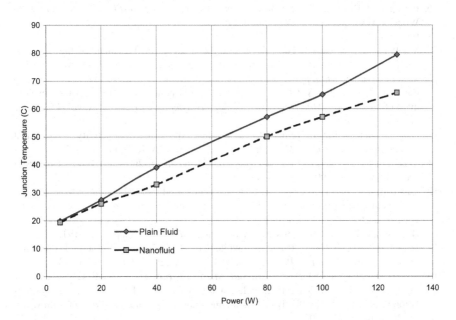

FIGURE 8.16 Average junction temperature of die A2 versus coolant flow rate for both plain fluid and ternary nanofluid. Both results used the all-metal nanoscale thermal interface.

ACKNOWLEDGEMENT

Sincere thanks go to Nick Teneketges and Tarzen Kwok from Teradyne Inc. for use of MV-1 test vehicles and setup, Mohsen Esmailpour for general support, and Andy Pfhanl for technical advice and help with the write-up.

REFERENCES

1. Simons, R.E., and K.P. Moran, "Immersion cooling systems for high density electronic packages," *Nepcon-West Proceedings*, Anaheim, CA, February 1977.
2. Mudawar, I., and T.M. Anderson, "Parametric investigation of the effects of coolant variation, pressurization, subcooling, and surface augmentation," *ASME-HTD(American Society of Mechanical Engineering—Heat Transfer Division)*—Vol. 111 Heat Transfer In Electronics, Book No. H00503, 1989, pp. 35–49.
3. Mudawar, I., and D.E. Maddox, "Critical heat flux in subcooled flow boiling of fluorocarbon liquid on a simulated electronic chip in a vertical rectangular channel," *Int. Journal of Heat Mass Transfer*, vol. 32, no. 2, 1989, pp. 379–394.
4. Mudawar, I., and D.E. Maddox, "Enhancement of critical heat flux from high power microelectronic heat sources in a flow channel," *ASME-HTD*—Vol. 111 Heat Transfer In Electronics, Book No. H00503, 1989, pp. 51–58.
5. Pfhanl, A., and J. Cepeda-Rizo, "Heat sinks reactively soldered to ICs," *DesignCon East, IEC (International Engineering Consortium) Conference Proceedings*, Boxborough, MA, April 2004, pp. 235–259.
6. Incropera, F.P., *Liquid Cooling of Electronic Devices by Single-Phase Convection*, John Wiley & Sons, New York, 1999, pp. 9–11.
7. Ibid., pp. 145–147.
8. Yu, W., and U.S. Choi, "The role of interfacial layers in the enhanced thermal conductivity of nanofluids: A renovated Maxwell model," *Journal of Nanoparticle Research*, vol. 5, 2003, pp. 167–171.
9. Eastman, J.A., and U.S. Choi, "Enhanced thermal conductivity through the development of nanofluids," *Materials Research Society Symposia Proceedings*, vol. 457, 1997, pp. 3–12.
10. Das, S.K., and N. Putra, "Temperature dependence of thermal conductivity enhancement for nanofluids," *Journal of Heat Transfer-ASME*, vol. 125, 2003, pp. 567–574.
11. Maxwell, J.C., *A Treatise on Electricity and Magnetism*, 2nd Ed., Oxford University Press, Cambridge, 1904, pp. 435–441.
12. Eastman, J.A., and U.S. Choi, "Anomalously increased thermal conductivities of ethylene glycol-based nanofluids containing copper nanoparticles," *Applied Physics Letters*, vol. 78, no. 6, 2001, pp. 718–720.
13. Hamilton, R.L., and O.K. Crosser, *I & EC Fundamentals*, 1962, pp. 187–191.
14. Jang, S.P., and U.S. Choi, "Role of Brownian motion in the enhanced thermal conductivity of nanofluids," *Applied Physics Letters*, vol. 84, no. 21, 2004, pp. 4316–4318.
15. *Mathis TC-01 System Operation Manual*, Mathis Instruments, Ltd, NB, Canada, 2005.
16. Bhattacharya, P., S.K. Saha, P.E. Phelan, and R.S. Prasher, "Brownian dynamics simulation to determine the effective thermal conductivity," *Journal of Applied Physics*, vol. 11, 2004, pp. 6492–6505.

9 Power Systems
The Tesla Turbine

INTRODUCTION

The gas was assumed to be isentropic so that the pressure and density could be related by a simple nonlinear equation of state involving the polytropic index. This was supplemented by the equations of conservation of mass and momentum, assuming axisymmetric, in the region between two rotating disks in the Tesla turbine. The viscosity-dominated regime could be solved completely but provided no torque at leading order. Extending the calculations to small but finite Reynolds number would provide an expression for the torque at first order in Reynolds number. However, since the operating regime involves high Reynolds numbers, this model was not pursued, and we instead focused on the inertia-dominated regime. In the latter case, for high enough driving pressures, it was predicted that the flow becomes "choked" and a standing circular shock wave must form at some finite radius within the disks. Modeling this regime further can be fruitful and provide new mathematical results. At lower driving pressures, the base Euler flow could also be calculated without shock waves. To get the resulting torque and angular velocity of the disks, the effects of viscous boundary layers on the disks would need to be accounted for. Our team did consider the boundary layer problem; however, only for the incompressible limit where the core flow was obtained by solving the constant density Euler equations. At that point, by assuming that the boundary layers are thin and uniform, we were able to derive an expression for the torque with or without applied loads. To investigate the boundary layer structure in more detail, the team also applied the Karman momentum integral approach to the 2D flow problem and, upon making an assumption regarding the self-similar structure of the velocity profiles, was able to derive a second analytical result for the torque as a function of the applied load. Several of the approaches pursued by the team during the week are worthy of further research. Once single-phase flow in the Tesla turbine is well understood, the case of vapor-liquid flow can be reconsidered.

INTRODUCTION

Our work in the study group was to examine the flow between two adjacent disks in a Tesla turbine. The problem was presented to the study group by Juan Cepeda-Rizo of JPL. A standard turbine has blades that "catch" the flow. The blades have an important role in making the turbines function, but their presence means that there must be a single-phase flow. A two-phase flow (wet steam or vapor-liquid flow) tends to damage the blades, at least at high speed. In 1913, Nikola Tesla proposed a new kind of turbine that has no blades and thus can handle two-phase flow [1]. For our problem we were

DOI: 10.1201/9781003247005-9

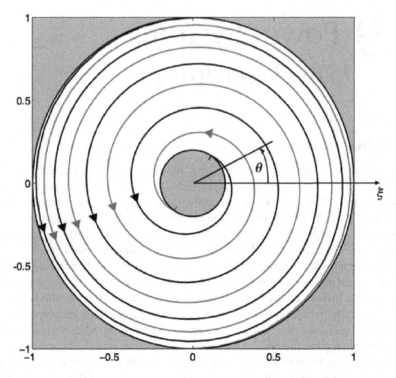

FIGURE 9.1 A schematic representation of the flow geometry. Steam follows the blue path in tangent to the outer edge of the disk, over the spinning disk, then out of the turbine through an outlet at the center.

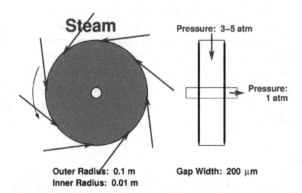

FIGURE 9.2 Schematic including the side view between two adjacent disks. The left-hand schematic indicates how typically there are multiple inlets arranged symmetrically around the turbine.

asked to examine two-phase flow in the Tesla turbine. But an examination of the literature indicated that little mathematical modeling work exists on single-phase flow in the Tesla turbine (cf., e.g., [2]). Therefore, we set out to examine single-fluid flow in the Tesla turbine. Schematics of the flow in a Tesla turbine are shown in Figures 9.1 and 9.2. As indicated in Figure 9.1, steam (or a water-steam mix) enters the turbine flowing tangent to a turbine disk, then spirals inward flowing over the disk. Initially, along the outer portion of the disk, the steam flows faster than the disk is spinning, but as the flow spirals inward, the flow slows and eventually exits the turbine through an outlet at the center of the disk. A side view of this flow is given in Figure 9.2. As is indicated on the figures, the outer radius of the disk is typically 10 cm, the inner radius (the radius of the flow exit) is typically 1 cm, and the gap width between disks is 200 μm. The inlet pressure of the flow near the outer edge of the turbine disks is typically 3–5 atm, while the outlet pressure near the center of the disks is approximately 1 atm. A number of recent articles, theses, and reports deal with various aspects of Tesla turbines; among these are [3–6]. This report is divided into three main sections: the next two sections discuss incompressible and compressible models of the flow, ignoring viscous effects. The final section then gives a boundary layer analysis. For the compressible case, the gas is assumed to be isentropic so that the pressure and density could be related by a simple nonlinear equation of state involving the polytropic index. This was supplemented by the equations of conservation of mass and momentum, assuming axisymmetry, in the region between two rotating disks in the Tesla turbine. The viscosity-dominated regime could be solved completely but provided no torque at leading order. Extending the calculations to a small but finite Reynolds number would provide an expression for the torque at first order in the Reynolds number. However, since the operating regime involves high Reynolds numbers, this model was not pursued and we instead focused on the inertia-dominated regime. In the latter case, for high enough driving pressures, it was predicted that the flow becomes "choked" and a standing circular shock wave must form at some finite radius within the disks. Modeling this regime further can be fruitful and provide new mathematical results. At lower driving pressures, the base Euler flow could also be calculated without shock waves. To get the resulting torque and angular velocity of the disks, the effects of viscous boundary layers on the disks would need to be accounted for. For the incompressible limit where the core flow was obtained by solving the constant density Euler equations, we considered the boundary layer problem. By assuming that the boundary layers are thin and uniform, we were able to derive an expression for the torque with or without applied loads. To investigate the boundary layer structure in more detail, we also applied the Karman momentum integral approach to the 2D flow problem and, upon making an assumption regarding the self-similar structure of the velocity profiles, were able to derive a second analytical result for the torque as a function of the applied load. Several of the approaches pursued by the team during the week are worthy of further research.

INCOMPRESSIBLE INVISCID MODEL

For the incompressible, inviscid case, the continuity and Navier-Stokes equations in cylindrical coordinates are used:

$$\frac{\partial}{\partial r}\left(\rho_0 r v_r\right) = 0,$$

(9.1)

$$\rho_0\left(v_r\frac{\partial v_r}{\partial r} - \frac{v_\theta^2}{r}\right) = \frac{\partial \rho}{\partial r},$$

(9.2)

$$\frac{\partial v_\theta}{\partial r} + \frac{v_\theta}{r} = 0.$$

(9.3)

In these equations, terms that are deemed small by a scaling argument are dropped. We are also assuming that vz = 0, so the z-equation for Navier-Stokes in cylindrical coordinates does not appear in this analysis. The boundary conditions on v_r and v_θ at the edge of the disc, corresponding to r = R, are set to v $_r$ ^{R} and v R θ, respectively. Integrating Equations (9.1) and (9.3) in r and applying the boundary conditions yields equations for v_r and v_θ:

$$v_r\left(r\right) = v_r^R\left(\frac{R}{r}\right),$$

(9.4)

$$v_\theta\left(r\right) = v_\theta^R\left(\frac{R}{r}\right).$$

(9.5)

With these two velocities, we can integrate Equation (9.2) in r to solve for the pressure:

$$p\left(r\right) = -\frac{1}{2}\rho_0\left|v_{in}\right|^2\left(\frac{R}{r}\right)^2,$$

(9.6)

$$\left|v_{in}\right| = \left(v_r^R\right)^2 + \left(v_\theta^R\right)^2.$$

(9.7)

The change in pressure from the outer radius, R, to the inner radius εR, is defined as p(R) − p(εR) = Δp. The parameter ε is the ratio of the inner radius to the outer radius. With the pressure now defined, we can solve for the radial velocity at the edge of the disc, v R r, in terms of the angular velocity at the edge of the disc:

$$v_r^R = -\sqrt{\frac{2\Delta p}{\rho_0}\left(\frac{\varepsilon^2}{1-\varepsilon^2}\right) - \left(v_\theta^R\right)^2}.$$

(9.8)

This is valid provided the value under the square root remains nonnegative, or in terms of the change in pressure, Δp:

$$\Delta p > \frac{\rho_0\left(v_\theta^R\right)^2}{2\varepsilon^2}\left(1-\varepsilon^2\right).$$

(9.9)

The shear stress on each disc, where δ is the boundary layer thickness (assumed uniform in this section) and Ω is an angular velocity of the disc, is given by:

$$\tau_\theta = \frac{\mu}{\delta}\left(v_\theta(r) - V_\theta(r)\right), \tag{9.10}$$

$$V_\theta(r) = \Omega r. \tag{9.11}$$

The torque on each disk is then:

$$T = \frac{\pi \mu R^3}{\delta}\left[v_\theta^R\left(1-\varepsilon^2\right) - \frac{\Omega R}{2}\left(1-\varepsilon^4\right)\right]. \tag{9.12}$$

We assume that there is no external load, meaning that $T = 0$. With this assumption, the angular velocity is:

$$\Omega = \frac{2v_\theta^R}{\left(1+\varepsilon^2\right)R}. \tag{9.13}$$

The mass-flow rate is given by:

$$\dot{m} = -4\pi R h \rho_0 v_r^R, \tag{9.14}$$

where h is the half of the distance between the two disks. Using the previously derived form of v R r, Equation (9.8), the mass-flow rate is:

$$\dot{m} = 4\pi R h \rho_0 \sqrt{\frac{2\Delta p}{\rho_0}\left(\frac{\varepsilon^2}{1-\varepsilon^2}\right) - \left(v_\theta^R\right)^2}. \tag{9.15}$$

We can solve Equation (9.15) for the angular velocity:

$$\Omega = \frac{2}{\left(1+\varepsilon^2\right)R}\sqrt{\frac{2\Delta p}{\rho_0}\left(\frac{\varepsilon^2}{1-\varepsilon^2}\right) - \left(\frac{\dot{m}}{4\pi R h \rho_0}\right)^2}. \tag{9.16}$$

The parameters for air at 20 degrees Celsius provide an upper bound (at 1 atmosphere) for the angular velocity, Ω:

$$\Omega_{max} = 800\,\text{rad/s} \approx 8000\,\text{rpm}. \tag{9.17}$$

This derivation assumes that the boundary layer, δ, is constant across the disc, but we can also consider the case when it is not.

COMPRESSIBLE INVISCID MODEL

Since in turbines gas enters at high pressures and exits at much lower pressures, compressibility effects often need to be accounted for. For typical axial flow turbomachinery, 1D compressible flow models are used to describe the flow and, depending on the circumstances (e.g., when cross-sectional area of a compressor or turbine changes), there is a possibility that one obtains "choked" flow, with standing shock waves forming somewhere within the turbomachine. For the Tesla turbine, the cross-sectional area available to the flow decreases as the fluid flows radially inward, so it appears that similar choked flow conditions may be obtained in this geometry. Our modeling seems to support this idea, though further analysis is certainly warranted. Again, this section assumes that the flow is steady state (time independent), that it is axisymmetric (independent of θ), and that $v_z = 0$. For compressible, inviscid flow, the continuity (9.18) and Euler (9.19)–(9.21) equations (in cylindrical coordinates) are given next, along with the isentropic state equation (9.22), which relate the density and pressure (and also captures the energy equation):

$$\frac{1}{r}\frac{\partial}{\partial r}\left(\rho r v_r\right) = 0, \tag{9.18}$$

$$\rho\left(v_r\frac{\partial v_r}{\partial r} - \frac{v_\theta^2}{r}\right) = -\frac{\partial p}{\partial r}, \tag{9.19}$$

$$\rho\left(v_r\frac{\partial v_\theta}{\partial r} + \frac{v_\theta v_r}{r}\right) = 0, \tag{9.20}$$

$$0 = -\frac{\partial p}{\partial z}, \tag{9.21}$$

$$\frac{p}{p_0} = \left(\frac{\rho}{\rho_0}\right)^\gamma. \tag{9.22}$$

The boundary conditions on v_r and v_θ at $r = R$ are again $v_r\, r = R = v\, R\, r\, (< 0)$ and $v_\theta\, r = R = v\, R\, \theta$. For pressure, the boundary conditions are $p\, r = R = p_0 + \Delta p$, and $p\, r = R = p_0$. Equation (9.21) implies that p and thus ρ are functions of r alone. Equation (9.20) and the boundary condition for v_θ imply that $v_\theta(r) = v\, R\, \theta\, R/r$. Also the continuity equation (9.18) integrates immediately to yield that $\rho(r)v_r(r) = \rho(R)v\, R\, r\, R/r$. Since the main factor that determines whether choked flow conditions occur is the change in total cross-sectional area available to the flow, the θ component of the flow is not as essential in this analysis while the radial flow is quite important. Therefore, let us consider a purely radial flow, neglecting the angular component of the velocity for the time being. Substituting the expression for v_r into Equation (9.19) and using the equation of state (9.22), one finds a single ordinary differential equation for:

$$p : dp \; dr = C2\rho 0r2 \; p0 \, p1 \gamma 1r + 1\gamma \, p \; dp \; dr \qquad (9.23)$$

where $C = \rho(R)v \; R \; r \; R$. Upon rendering this equation dimensionless, a simpler form of the ODE can be obtained:

$$dP \; dX = \gamma PX \left(\gamma AX2P \alpha - 1 \right) \qquad (9.24)$$

where $P \equiv p/p0$, $X \equiv r/R$, $A \equiv \rho 0p0R2 \; C2$ and $\alpha \equiv 1+ 1 \; \gamma$. The corresponding boundary conditions are:

$$P(1) = 1 + \Delta p \; p0 \; , \; P() = 1 \qquad (9.25)$$

Theorem 1 *The pressure equation has no smooth solution for* $\dfrac{\Delta p}{p_0} > \exp \sqrt{\dfrac{-2\gamma \ln \varepsilon}{\alpha}}$

This nonlinear differential equation was studied during the workshop and one key result that emerged, consistent with the possibility that choked flow conditions occur and a shock wave forms, was obtained as follows. Analysis of Equations (9.24) and (9.25) suggests that when the pressure difference between the inlet and outlet is sufficiently large, no smooth solution of (9.24) is possible that satisfies. The pressure equation has no smooth solution for $\Delta p \; p0 > \exp r \; -2\gamma \ln \alpha$. This result implies that the solution must be discontinuous, and therefore the flow must have a shock. Such a flow is said to be choked.

Boundary Layer (BL) Analysis

Detailed analysis of the boundary layer structure on the rotating disks with the inviscid core in between them turned out to be too complex to be completed during the brief period of the workshop. We thus decided to apply Karman's momentum integral approximation with the hope of obtaining an approximate expression for the dependence of the BL thickness on the radial distance from the edge of the disks, together with corresponding approximations for the shear stress and torque on each disk.

4.1 Momentum Integral Approach

Karman's momentum integral approach starts with the BL equations $1 \; r \; \partial(rvr) \; \partial r + \partial vz \; \partial z = 0$ (9.26), $vr \; \partial vr \; \partial r - v \; 2 \; \theta \; r + vz \; \partial vr \; \partial z = -1 \; \rho \; dp\infty \; dr + \mu \; \rho \; \partial \; 2 \; vr \; \partial z2$ (9.27), and $vr \; \partial v\theta \; \partial r + v\theta vr \; r + vz \; \partial v+ \; \partial z = \mu \; \rho \; \partial \; 2 \; v\theta \; \partial z2$ (9.28) in which, in addition to the previous terms appearing in the Euler equations, the dominant terms from the viscous contributions in the Navier-Stokes equations are retained. Namely, the terms involving the second derivatives with respect to z on the right-hand sides of the Equations (9.27) and (9.28) are expected to be the dominant viscous terms. This implies that the velocity components also depend upon z (vertical distance from a disk surface), although the pressure is still uniform across the BL and thus independent of z. The momentum integral approach involves integrating all three of the previous equations with respect to z from 0 to some chosen height H well above the BL. After the required manipulations are complete, one can take the limit H $\rightarrow \infty$ to complete the derivation. It should be noted that when coordinate z is in the region outside the BL, all z-derivatives vanish and vr $\rightarrow$ v ∞ r (r) and vθ $\rightarrow$ v ∞ θ (r), approaching their core inviscid limits, obtained earlier. Integration of the continuity equation in this manner yields:

$$vz(r, H) + 1 \, r \, d \, dr \, r \, Z \, H \, 0 \, vr(r, z)dz = 0 \tag{9.29}$$

Integration of the r- and θ-momentum equations with respect to z, use of integration by-parts, simplification using the integrated form (9.29) of the continuity equation, and taking the limit as H $\rightarrow \infty$ result in:

$$\frac{1}{r}\frac{d}{dr}\left[r\int_0^\infty v_r\left(v_r^\infty - v_r\right)dz\right] + \frac{dv_r^\infty}{dr}\int_0^\infty \left(v_r^\infty - v_r\right)dz - \frac{1}{r}\int_0^\infty \left[\left(v_\theta^\infty\right)^2 - v_\theta^2\right]dz = \frac{\tau_r}{\rho}, \tag{9.30}$$

$$\frac{1}{r^2}\frac{d}{dr}\left[r^2\int_0^\infty v_r\left(v_\theta^\infty - v_\theta\right)dz\right] = \frac{\tau_\theta}{\rho}. \tag{9.31}$$

In the last two equations, the shear stress components on the disk are defined as before: $\tau r = \mu \, \partial vr \, \partial z \, z=0 \, \tau\theta = \mu \, \partial v\theta \, \partial z \, z=0$ To proceed further, one assumes certain "similarity" profiles for the velocity fields. For instance, let us denote the r-dependent BL thickness by $\delta(r)$ and take, for $z \leq \delta(r)$:

$$v_r\left(r,z\right) = v_r^\infty \sin\left(\frac{\pi z}{2\delta}\right) \tag{9.32}$$

$$v_\theta\left(r,z\right) = \Omega r + \left[v_\theta^\infty - \Omega r\right]\sin\left(\frac{\pi z}{2\delta}\right). \tag{9.33}$$

This is analogous to one of the common approximations made in standard BL theory for flow past a flat plate. Note that when z = 0, the velocity components match those on the rotating disk (i.e., the no-slip boundary condition is satisfied), and when z = $\delta(r)$, i.e., at the top of the BL, the velocities match the ones in the core inviscid flow. However, the z-variation in both velocity components would be the same under this assumption. Alternatively, one can also take:

$$v_\theta\left(r,z\right) = \Omega r\cos\left(\frac{\pi z}{2\delta}\right) + v_\theta^\infty \sin\left(\frac{\pi z}{2\delta}\right) \tag{9.34}$$

which represents a different interpolation of the θ velocity between the disk and core. Substitution of (9.32) and (9.33) into the momentum integral equations (9.30) and (9.31) results in two separate first order ODEs for $\delta(r)$ with Ω as an unknown parameter. The equations are not reproduced here (they are easy to derive) since it does not appear that a single value of Ω will make the two ODEs consistent with each other. Thus, the first approximation does not appear to yield a consistent solution. We can, nevertheless, impose the condition that the disk be torque-free and obtain a useful result. Recall that the torque-free requirement is tantamount to:

$$\int_0^{2\pi} \int_{\varepsilon R}^{R} \left(r\tau_\theta \right) r\, dr\, d\theta = 0. \tag{9.35}$$

Substitute the expression for the θ component of the shear stress from the second momentum integral equation (9.31) into the torque-free condition to find:

$$\int_0^{\infty} v_r \left(v_\theta^{\infty} - v_\theta \right) dz = 0 \quad at \quad r = \varepsilon R \tag{9.36}$$

This assumes that the BL thickness starts out at a value of zero at the edge of the disks $r = R$, which is the inflow. Using the interpolation (9.32) for vr and the alternative interpolation of vθ given by (9.34) in the torque-free condition.

This provides a relationship between the angular velocity of the disk and the θ component of the core flow velocity at the inner radius of the Tesla turbine disks. The Karman momentum integral approach appears to offer a promising track for obtaining an approximate structure for the r-dependent BL thickness. It would seem particularly worthwhile to use the velocity interpolations (9.32) and (9.34) in the two momentum integral equations (9.30) and (9.31) to see if a consistent pair of differential equations can be obtained for the BL thickness $\delta(r)$ when the parameter Ω is chosen so as to satisfy the torque-free condition (or to match an imposed torque "load" on the turbine).

FUTURE WORK

The work described previously was carried out in three days during January 2011. Of course, there are many other issues that should be considered, including the original question brought to the workshop: that of two-phase flow and condensation in the Tesla turbine. A number of other avenues for future work also might be fruitfully pursued:

- complete inviscid solution for small pressure gradients
- associated compressible boundary layer analysisx
- include thermal effects and implication of shocks on the efficiency
- more detailed boundary layer analysis via the momentum integral approach
- analyze the viscous-dominated regime for smaller gap sizes.

ARRIVING AT A CLOSED-FORM SOLUTION

In addition to the laminar flow assumption, the following idealizations are adopted here.

(1) The flow is taken to be two-dimensional: $vz = 0$, vr, and v are taken to be constant across the channel between rotor in Equation plates (9.2), and hence, are mean flow velocities at each r and (2) the flow field with vr and v velocity components is treated as being inviscid, with a body-force representation of the wall shear effects.

The viscous drag exerted on the flow by the sidewalls of the channel between the rotors is modeled as a body force acting on the flow at each r location. (3) The flow field is radially symmetric. The inlet flow at the rotor outer edge is uniform, resulting in a flow field that is the same at any angle. All derivatives of flow quantities are therefore zero. (4) Radial pressure gradient effects are negligible, as compared to angular momentum and wall frictional drag effects. With these idealizations, the previous four governing equations reduce to (5).

Continuity:

$$\frac{1}{r}\frac{\partial(rv_r)}{\partial r}+\frac{1}{r}\frac{\partial v_\theta}{\partial \theta}+\frac{\partial v_z}{\partial z}=0 \tag{9.1}$$

r-direction momentum:

$$v_r\frac{\partial v_r}{\partial r}+\frac{v_\theta}{r}\frac{\partial v_r}{\partial \theta}+v_z\frac{\partial v_r}{\partial z}-\frac{v_\theta^2}{r}=-\frac{1}{\rho}\left(\frac{\partial P}{\partial r}\right)+$$

$$v\left\{\frac{1}{r}\frac{\partial}{\partial r}\left(r\frac{\partial v_r}{\partial r}\right)+\frac{1}{r^2}\frac{\partial^2 v_r}{\partial \theta^2}+\frac{\partial^2 v_r}{\partial z^2}-\frac{v_r}{r^2}-\frac{2}{r^2}\frac{\partial v_\theta}{\partial \theta}\right\}+f_r \tag{9.37}$$

θ-direction momentum:

$$v_r\frac{\partial v_\theta}{\partial r}+\frac{v_\theta}{r}\frac{\partial v_\theta}{\partial \theta}+v_z\frac{\partial v_\theta}{\partial z}+\frac{v_r v_\theta}{r}=-\frac{1}{\rho}\left(\frac{\partial P}{\partial \theta}\right)+$$

$$v\left\{\frac{1}{r}\frac{\partial}{\partial r}\left(r\frac{\partial v_\theta}{\partial r}\right)+\frac{1}{r^2}\frac{\partial^2 v_\theta}{\partial \theta^2}+\frac{\partial^2 v_\theta}{\partial z^2}-\frac{v_\theta}{r^2}-\frac{2}{r^2}\frac{\partial v_r}{\partial \theta}\right\}+f_\theta \tag{9.38}$$

z-direction momentum:

$$v_r\frac{\partial v_z}{\partial r}+\frac{v_\theta}{r}\frac{\partial v_z}{\partial \theta}+v_z\frac{\partial v_z}{\partial z}=-\frac{1}{\rho}\left(\frac{\partial P}{\partial z}\right)+$$

$$v\left\{\frac{1}{r}\frac{\partial}{\partial r}\left(r\frac{\partial v_z}{\partial r}\right)+\frac{1}{r^2}\frac{\partial^2 v_z}{\partial \theta^2}+\frac{\partial^2 v_z}{\partial z^2}\right\}+f_z \tag{9.39}$$

With these idealizations, the prior four governing equations reduce to:

$$\frac{1}{r}\frac{\partial(rv_r)}{\partial r}=0 \tag{9.40}$$

$$v_r\frac{\partial v_r}{\partial r}-\frac{v_\theta^2}{r}=-\frac{1}{\rho}\left(\frac{\partial P}{\partial r}\right)+f_r \tag{9.41}$$

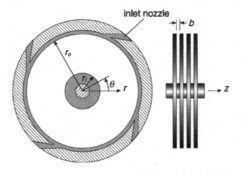

FIGURE 9.3 Tesla turbine coordinate system and dimensions.

$$\upsilon_r \frac{\partial \upsilon_\theta}{\partial r} + \frac{\upsilon_r \upsilon_\theta}{r} = f_\theta \tag{9.42}$$

$$0 = \frac{1}{\rho}\left(\frac{\partial P}{\partial z}\right) \tag{9.43}$$

$$\upsilon_r = -\frac{\dot{m}_c}{2\pi rb\rho} \tag{9.44}$$

$$F_\theta = \tau_w A_w = 4\tau_w V_e/D_H \tag{9.45}$$

Van Carey shows in his paper that the following closed form solution can calculated:

$$\eta_{rm} = \frac{\left(\hat{W}_0 + 1\right) - \left(\hat{W}_i + \xi_i\right)\xi_i\left(\gamma - 1\right)M_o^2}{\left[1 - \left(\frac{P_i}{P_{nt}}\right)^{(\gamma-1)/\gamma}\right]} \tag{9.46}$$

where ηrm is the rotor efficiency and the rotor diameter and other dimensions are contained with the parameters of the equation as shown in Table 9.1. A plot of Equation (9.45) as a function of rotor diameter is shown in Figure 9.4.

Where

η_{rm} is the rotor mechanical efficiency

$\hat{W}_0$ is the dimensionless variable of velocity

ξ rotor inner diameter to outer diameter ratio

γ specific heat ratio

$\frac{P_i}{P_{nt}}$ compression ratio

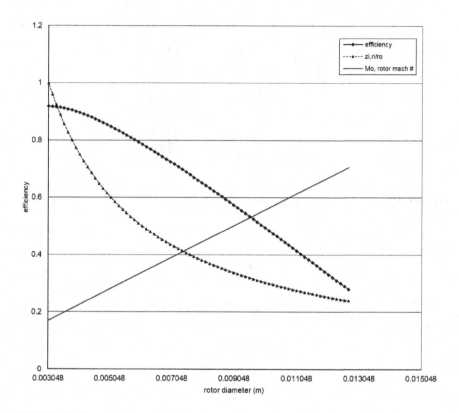

FIGURE 9.4 Plot of turbine efficiency vs. rotor diameter.

Now that we have the expected turbine efficiency versus the various parameters of the Tesla turbine, we could now estimate the cooling performance of the turbine in a reverse Brayton application.

$\dot{Q}_{net}$ = net cooling $\dot{P}_{DC,inv}$ = DC power into the compressor inverter T3 Heat Rejection state point T6 Cooling Load state point $\dot{W}_{net}$ = turbine refrigeration $\dot{Q}_{recup}$ = recuperator loss, $\dot{m}$ = cycle mass-flow rate, Cp = specific heat of cycle gas, ε = thermal effectiveness of the recuperator $\eta_{aero,t}$ = aerodynamic efficiency of the turbine (actual expansion power divided by isentropic expansion power) $\dot{m}_{leak,t}$ = flow rate of the cycle gas that bypasses the turbine $\dot{W}_{drag,t}$ = turbine drag, and $\dot{W}_{alt,t}$ = alternator losses.

CASE STUDY, AN AUTOMOTIVE AIR CONDITIONER

Chrysler Ltd. came to JPL in 2010 looking for innovative ideas to apply to the automotive industry. It was proposed to them to employ a Tesla turbine as a reverse Brayton cooler to provide cabin cooling of the automobile. An automotive cooler that received heat input from the engine exhaust and converted a portion of the energy as heat lift for cooling the cabin was chosen as a case study. Reverse Brayton

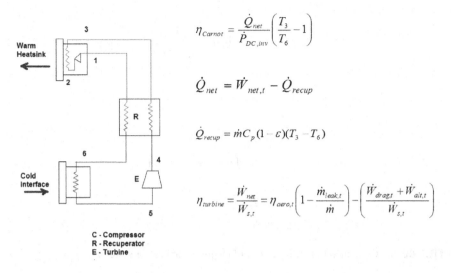

C - Compressor
R - Recuperator
E - Turbine

FIGURE 9.5 $\dot{Q}_{heatleak,t}$ = heat transfer from the warm to cold end of the turbine, $\eta_{aero,t}$ = ratio of isentropic to actual compressor power, $\dot{W}_{motor,c}$ = motor losses, and $\dot{W}_{inv,c}$ = inverter losses.

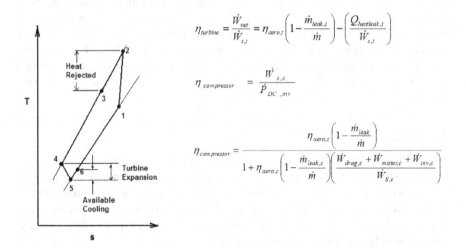

FIGURE 9.6 Calculated efficiency of turbine and compressor.

coolers have been used extensively in space applications by Creare for cryocooler applications.

A reverse Brayton cycle cooler that uses air as the refrigerant is proposed to replace mainstream R134a vapor compression systems in an automotive vehicle. The proposed system requires 1.3–1.7 kW from the engine, harnessed from the exhaust waste heat, and is capable of lifting 4.5–5.6 kW of heat (1.25–1.56 tons

$$\eta_{carnot} = \eta_{comp} \cdot \eta_{turbine} \left(1-\beta\right)\left(\dfrac{1}{PR^{\frac{\gamma-1}{\gamma}}}\right)\left(1-\dfrac{T_6}{T_3}\right)$$

$$\eta_{carnot} \propto \eta_{turbine}$$

$$\eta_{carnot} \propto \eta_{compressor}$$

$$\eta_{carnot} \propto \dfrac{1}{PR^{.287}}$$

$$\eta_{carnot} \propto \Delta T, \left(\Delta T = T_{hot} - T_{cold}\right)$$

$$\eta_{carnot} \propto \dfrac{1}{T_{hot}}$$

$$\eta_{carnot} \propto \beta$$

FIGURE 9.7 Combined Carnot efficiency of the turbocompressor system.

refrigeration). This is sufficient for providing air conditioning for a standard midsize vehicle under hot static conditions similar to a test parameters report submitted by Chrysler. A published study demonstrated that a system can be created with an off-the-shelf turbocharger that would be 56% more expensive and 10% heavier, but a bladeless turbocompressor is proposed to bring down both cost and weight as well as to increase the overall system efficiency and consume less fuel. Reverse Brayton systems can run either as an open or a closed system, and both have their own benefits. There is enough exhaust heat to run the system while in idle mode; though the system could easily be configured to be pulley driven at the cost of reduced efficiency.

REVERSE BRAYTON CYCLE

What is being proposed is a method of achieving refrigeration by utilizing reverse Brayton cycle cooling similar to what is used currently in commercial aircraft. NASA used a reverse Brayton cryocooler from Creare to cool the near infrared camera and multi-object spectrometer (NICMOS) instrument onboard the Hubble Space Telescope. Brayton cycle coolers are also commonly used in the condensation and production of liquid nitrogen from atmospheric air.

In the proposed embodiment (Figures 9.1 and 9.2), exhaust combustion gas enters the turbocompressor on the hot side and spins the turbine, which is coupled to the cold-side compressor via a shaft. Ambient air at State 1 enters the compressor and compresses the air to State 2. The hot and compressed gas now enters the isobaric intercooler and cools the air to near ambient temperature at State 3. The air enters the third turbine and undergoes isentropic expansion to a cold State 4. A turbo alternator converts the rotation energy of the expander into electricity, which is used to charge the vehicle's electrical system battery or a hybrid battery for auxiliary power.

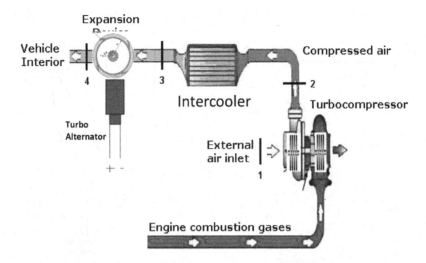

FIGURE 9.8 Proposed reverse Brayton cycle automotive A/C system.

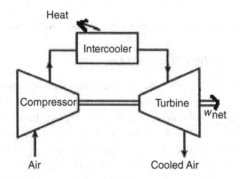

FIGURE 9.9 Standard diagram depicting an open-cycle reverse Brayton A/C.

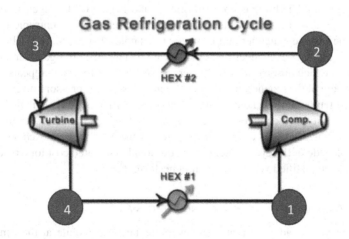

FIGURE 9.10 Closed-cycle reverse Brayton A/C.

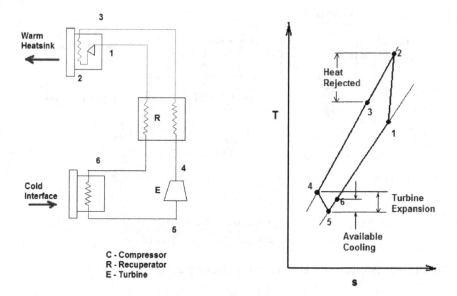

FIGURE 9.11 Closed-cycle reverse Brayton with regeneration (Creare, Inc.).

CLOSED-CYCLE SYSTEM WITH REGENERATION

Another approach for gain in efficiency is to create a simple closed-cycle system that recirculates dry air (Figure 9.10). Exhaust gas can still be used to power the compressor, but now the intercooler is replaced by two heat exchangers; a system like this would look very similar to the standard vapor compression systems that have one heat exchanger at the condenser and one at the evaporator. A more efficient approach would be to use a third regenerative heat exchanger (Figure 9.11). Dry air enters the compressor at State 1 and through isentropic compression heats to State 2, which is run through a heat exchanger similar to a condenser heat exchanger in a vapor compression system. The hot gas cools to State 3 and is sent to the regenerator, which cools the gas even further to State 4. The gas then undergoes isentropic expansion through the turbine expander and is cooled to its coldest state, State 5. The cold gas enters the cold heat exchanger and enters State 6. The still-cold gas then enters the regenerator, which absorbs heat and enters State 1, where the cycle repeats.

The benefit of the regenerative closed system is lesser compressor power to achieve the cooling temperatures and thus improved overall efficiency. Also in all closed systems, the quality of the air refrigerant is controlled, which increases the life span of the system. The drawback over an open system is the need for heat exchangers at the hot and cold side of the cycle. There is an additional associated cost for a regenerator, which must be justified versus the gain in performance.

BASELINE AIR-CONDITIONER SIZING AND PERFORMANCE

A baseline case is adopted that assumes a standard automobile at the time of the study (1990 Pontiac Grand Prix), an ambient temperature of 52 °C (125 °F), an in-car

TABLE 9.1
Baseline Air-Conditioning and Vehicle Parameters

Ambient temperature	52 °C
In car temperature	24 °C
Solar irradiation	900 W/m²
Driving speed	55 mph
Air density at 52 °C	1.1 kg/m3
Air density at 24 °C	1.3 kg/m3
Thermal conductivity of plate glass	1.4 W/mK
Windshield thickness	0.25 in
Area of windshield ('90 Pontiac Grand Prix)	1.4 m²
Total window area ('90 Pontiac Grand Prix)	3.41 m²
Max. A/C Pull-down time	120 sec

temperature of 24 °C (75 °F), a vehicle speed of 55 mph and a maximum hot soak pull-down time of 120 seconds [1].

AMBIENT CONDITIONS AND CONSTANTS

HEAT LOAD CHARACTERISTICS

The vehicle heat load is dominated by two large heat sources: solar irradiation and body conductance. It is a surprising result that when the car is moving, the largest single heat load on the system is not solar but rather heat conduction/convection through the vehicle windows. In other words, in hot weather driving, the faster the vehicle travels the more heat convection intrusion dominates over solar radiation.

SOLAR IRRADIATION

Solar radiation is the first major source of heat input to a passenger vehicle, and at the surface of the earth is assumed to be 900 W/m². The total window area of the 1990 Pontiac Grand Prix is 3.41 m², but realizing that half of the area is exposed to the sun at any given time, the maximum is reduced to 1.7 m². When a solar view of 30 degrees is taken into account, this area reduces to 1.5 m². Glass transmittance is 1.00. With this information the solar load is calculated as 1.35 kW.

TABLE 9.2
Baseline Solar Load

Solar irradiation	900 W/m²
Total window area	3.4 m²
Effective window area	1.5 m²
Solar load	**1.35 kW (thermal)**

TABLE 9.3
Baseline Body Conductance Load

Conductance	80 W/K
Temperature difference	28 °C
Body load	**2.24 kW**

Body Conductance

Though solar radiation heats the vehicle and passenger compartment during the day whether or not the vehicle is running, most air conditioning use occurs when the vehicle is in motion. We now consider heat that is brought into the vehicle by air convection, then by conduction through the body panels and windows of the vehicle. For the vehicle velocities considered, the body conductance roughly varies linearly with velocity. When traveling at 55 mph, this body conductance has a typical value of 80 W/K. When the temperature difference between the outside and inside of the car is 28 °C (52 °C – 24 °C), the body conductance is 2.24 kW, as shown in Table 9.3. This is a rough estimate that shows that heat intrusion by conduction into the vehicle roughly varies linearly with the speed of the vehicle and temperature difference with the outside environment. Table 9.3 shows that this body load is 60% larger than the solar load.

Other Loads

To sum up the steady-state heat loads affecting the passenger space, we must also account for outside air and passenger heat loads. The most important of these external loads is the amount of outside air, at ambient temperature, being brought into the passenger compartment. This is a variable load that can be adjusted by the driver of the vehicle. One extreme case involves full inside air recirculation with no replacement, while the other extreme replaces the air about twice a minute. A conservative amount of replacement air, 0.25 m^3/min, is assumed for baseline case. The calculation for the replacement air cooling requirement at 0.25 m^3/min and delta T of 28 °C gives 0.12 kW. The thermal impact of the replacement air is higher for higher flow rates.

In addition, the driver and passenger give off heat, assumed 120 W of heat for a total of 0.24 kW. Finally a small amount of heat leaks in from the engine and transaxle, assumed to total 0.50 kW of heat input involves full inside air recirculation with no replacement, while the other extreme replaces the air about twice a minute. A conservative amount of replacement air, 0.25 m^3/min, is assumed for baseline case. The calculation for the replacement air cooling requirement at 0.25 m^3/min and delta T of 28 °C gives 0.12 kW. The thermal impact of the replacement air is higher for higher flow rates.

In addition, the driver and passenger give off heat, assumed 120 W of power for a total of 0.24 kW. Finally, a small amount of heat leaks in from the engine and

TABLE 9.4
Total Baseline Heat Load

		Fraction
Solar radiative load at 900 W/m²	1.35 kW	0.30
Body conductance load at 55 mph	2.24 kW	0.50
Recirculated air energy	0.12 kW	0.03
Driver + one passenger	0.24 kW	.06
Engine loads	0.50 kW	0.11
Total heat load	**4.50 kW**	**1.00**

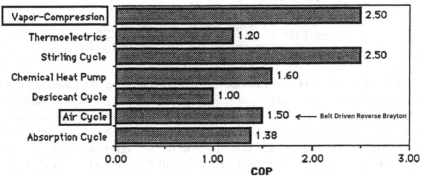

FIGURE 9.12 Automotive A/C technology comparison.

transaxle, assumed to total 0.50 kW of heat input. The steady-state heat load summary is given in Table 9.4.

EFFICIENCY COMPARISON: GAS REFRIGERATION (REVERSE BRAYTON) VS. VAPOR COMPRESSION (RANKINE)

A commonly used method for comparing the efficiency of two A/C systems is the coefficient of performance (COP), which is defined as the energy rate of cooling/input energy. The COP index gives unfair advantage to systems that use electric power for the input energy and is not a fair comparison with systems that derive input from waste heat. If we decided not to use waste heat to power the compressor and used a belt drive to run the compressor, the technologies would compare as seen in Figure 9.12.

As can be noticed in the figure, R134a vapor compression has a higher COP efficiency than does the "air cycle," which is also known as the belt-driven reverse Brayton cycle A/C. For the proposed system deriving its input power from automotive exhaust gas, the COP values must be normalized to meet the following definition:

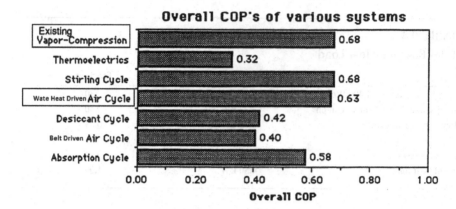

FIGURE 9.13 Normalized automotive A/C technology comparison.

$$\text{overall } COP = \frac{heat\_removed\_from\_vehicle}{heating\_value\_of\_required\_fuel}$$

The normalization of COP is accomplished by assuming that the base energy efficiency of the automobile engine is 30%, leaving 70% of the fuel energy as radiator or exhaust waste. For a waste heat-driven A/C system, the heat exchanger is assumed to recover 60%, resulting in an overall 42% efficiency. For the belt-driven or electrically driven systems, the drive system is assumed to be 90% efficient, which combined with the base engine efficiency results in an overall 27% efficiency. These normalized factors are multiplied by the basic COP to give an overall COP.

PERFORMANCE PARAMETERS

An ideal reverse Brayton cycle analysis was performed to show such a system is capable of removing the necessary 4.5 kW vehicle heat load by using the waste heat of the exhaust. Assuming a pressure ratio of 2.25 at the compressor from the four states shown previously in Figure 9.12, the analysis resulted in the following temperatures:

T1 = 10 °C   h1 = 283.15 kJ/kg
T2 = 83.5 °C   h2 = 356.6 kJ/kg
T3 = 38 °C   h3 = 311.25 kJ/kg
T4 = −26.5 °C   h4 = 246.64 kJ/kg
Mass flow at compressor, mdot = 0.1285 kg/s (16.4 lbm/min)
Heat lift, q_L = (h1—h4)*mdot = 36.51*.1285 = **4.69 kW**

BLADELESS TURBINE TECHNOLOGY (E.G., TESLA TURBINE)

The simplicity of a Tesla-type turbine can allow a low-cost, reliable design for a turbocompressor and turbine expander that could be an attractive option for reverse

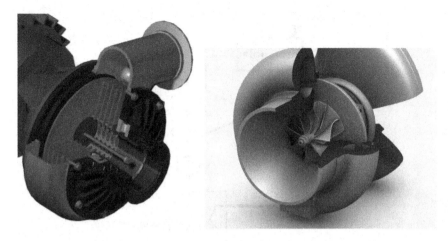

FIGURE 9.14 Tesla turbine compressor blades (left) vs. standard impeller blade compressor (right).

TABLE 9.5

Initial Costs for Conventional and Proposed AC Systems

Conventional System (R134a vapor compression)	
Main components	Cost [$]
Compressor	800.00
Evaporator/Expansion Device	500.00
Condenser	400.00
Total Initial Cost	**1700.00**
Proposed system (Rev. Brayton) with COTS turbocompressor	
Main components	Cost [$]
Turbocompressor	2,000.00
Intercooler	300.00
Evaporative System/Dryer	200.00
Expansion Device	150.00
Total Initial Cost	**2650.00**

Brayton air conditioning systems if an efficient design can be achieved. Though standard off-the-shelf turbocompressors (aka turbochargers or superchargers) can be designed into a reverse Brayton system, it is believed that the cost of the turbocompressors would make it unattractive compared to a standard vapor compression system [2].

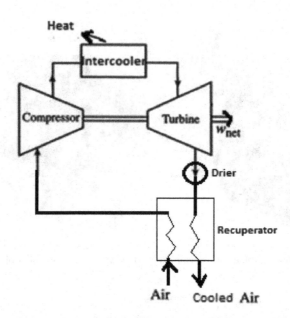

FIGURE 9.15 Recuperator added as a means of dehumidifying the air.

Cost Comparison

A study performed by Beatrice and Fiorelli demonstrated that the use of an existing commercial off-the-shelf (COTS) turbocompressor from Honeywell/Garrett would result in a system that is 40% more expensive than a standard vapor compression system (see Table 9.5) [3].

As can be seen from Figure 9.13, our reverse Brayton A/C system driven by waste heat has an overall COP of 0.63 compared to existing R134a vapor compression of 0.68. We believe that use of the bladeless turbine technology at the compressor and expander will increase the COP and even surpass that of vapor compression.

Available Waste Heat During Engine Idle

It is estimated that on average a midsize vehicle will consume about one gallon of gasoline for every 1 hour that it idles, which equates to 39 kW of wated heat energy. Considering experimental data provided by manufacturers, the turbocompressor increases the outlet engine pressure by 3.5% in the worst-case scenario, and the impact on the engine power is nearly the same. So if the engine power at idle is 39 kW, the turbocompressor impact is 1.35 kW, which is enough power to provide a decent amount of air conditioning and in the neighborhood of the input power required for a conventional air conditioning system.

Weight Comparison

By using off-the-shelf components, Beatrice and Fiorelli had their system evaluated at 11.5 kg, while a conventional vapor compression system weighed 10.5 kg. The turbocompressor accounts for over 50% of the total weight, similar to a conventional vapor compression system, and most of the weight is due to a very thick and heavy housing needed to meet safety standards from high-speed rotation. A Tesla turbine would not require such a thick housing because of the lower rotation inertia compared to centrifugal blades. It is estimated that the total weight of a system with a Tesla turbine should drop below 10 kg.

Dehumidification and Drying

As seen from the performance estimates of the system, the air will cool well below dew point and the moisture will have to be removed before being sent to the vehicle cabin. Off-the-shelf air dryers/desiccants exist that can be put into the system; they are conch shaped and remove dew from the air by rotating it and having it condense on the walls.

In the event that the air is not cooled substantially below dew point, we can place a recuperator between air intake of the compressor and air exit into the vehicle cabin. The recuperator lowers the temperature of the air as it enters the compressor, which allows for higher compressibility as colder air compresses more efficiently, but it also lowers the overall temperature of the air stream. As the air leaves the turbine expander, it drops lower than it would without the recuperator and the condensate can be collected by a dryer. As the air leaves the expander and enters the recuperator, it then heats the air back up, making it more comfortable for the passengers.

A third and simplistic approach would be to use membranes to dehumidify the air before it enters the Brayton unit; JPL has expertise in this area.

Final Recommendations

We would like to explore the use of a reverse Brayton cycle air conditioner as a means of replacing the standard R134a vapor compression units that are used today. A more detailed analysis can be performed to ascertain if a waste heat system is more advantageous than a pulley-driven system. To get the cost of such a system down into state-of-the-art vapor compression system range, we propose using a proprietary bladeless turbocompression system to provide power from exhaust heat, compression, and expansion of air, but would require non-recurring engineering (NRE) and development costs to produce. A second approach would be to use COTS components and attempt cost reduction by supply chain and procurement strategies.

REFERENCES

1. Fehribach, J., M. Mata, A. Nadim, J. Cepeda-Rizo, M. Gratton, and S.L. Smith, "Mathematical modeling of flow within a Tesla turbine," *AIM (American Institute of Mathematics) Sustainability Workshop*, Palo Alto, CA, 2011.

2. Carey, V., "Assessment of Tesla turbine performance for small scale Rankine combined heat and power systems," *Journal of Engineering for Gas Turbines and Power, ASME*, vol. 132, December 2010, pp. 1–8.
3. Beatrice, L., and F. Fiorelli, "Feasibility of a Brayton cycle automotive air conditioning system," *Engenharia Termica*, vol. 8, no. 2, December 2009; Engin, T., M. Ozdemir, and S. Cesmeci, "Design, testing and two-dimensional flow modeling of a multiple-disk fan," *Experimental Thermal and Fluid Science*, vol. 33, 2009, pp. 1180–1187.
4. Bloudcek, P., and D. Palousek, "Design of Tesla turbine," in *Konference diplomov´ych prac*, 2007; Harwood, P., "Further investigations into Tesla turbomachinery," *tech. rep.*, 2008. Supervisor: Prof. Mark Jones.
5. Ladino, A., "Numerical simulation of the flow field in a friction-type turbine (Tesla turbine)," PhD thesis, TU Wien, Vienna, Austria, 2004.
6. Tesla, N., Turbine, 1913. U.S. Patent No. 1061206; Multerer, B., and R.L. Burton, "Alternative technologies for automobile air conditioning," *Air Conditioning and Refrigeration Center*, University of Illinois Urbana, 1991.

10 Electronics Design for Extreme Temperature and Pressure

Venus missions have been a driver of high-temperature electronics for many years [1–2]. With a surface temperature of 450 °C and atmospheric pressure of 90 bar, a Venus surface would make a good testing ground for electronics capable of surviving extreme conditions. Many factors make operating electronics at room temperature on the surface not feasible, including the large mass associated with creating a suitable pressure vessel and the power required to cool it down some 400 °C would make the task daunting. Many of the common electronic packaging materials such as eutectic tin-lead solder, polymerics, wire insulators, and so on could not handle such high temperatures, so a start is to create a list of compatible materials or to eliminate their need all together. The JPL microelectronics division designed a robust solar cell to handle temperatures near 450 °C by carefully selecting the materials. Most IC base materials such as GaAS, GaN, SiO_2, and indium phosphide are well suited to withstand the high temperatures because they are ceramics. Most space-faring metals are also capable of surviving 450 °C. The solar cell for the Full Spectrum Power for Optical/Thermal eXergy (FSPOT-X) solar concentrator used a technique called hyper-CTE matching, where a suitable substrate to hold and house the solar cell was identified with less than 3E-06 μm/μm CTE matching between solar cell material (GaAs); this candidate material was Al_2O_3 gamma-aluminum oxide, with aluminum silicon-carbide (AlSiC) as a backup material.

The AlSiC material offers customization and good thermal conductivity, but it is very expensive and not well suited for the intent of the FSPOT program, which was to create a commercial and inexpensive solar concentrator.

The GaAs solar cell is well matched to both aluminum oxide and AlSiC substrates.

The photovoltaic (PV) cell was attached to the substrate with eutectic gold, which had the ability to melt at a temperature below 300 °C, with a remelt temperature of 515 °C.

For interconnects, as mentioned, the PV cell was attached to the substrate with eutectic gold, and the bus bar and interconnect of the three-junction design was wire bonded from the PV cell to metallized traces on the substrate. Since the package was leadless, gold ribbon was fusion bonded, or "ribbon bonded," using a friction ribbon bonder, thus avoiding an intermediate material such as tin-lead solder. The final result was a package tested to 370 °C, but it was perfectly capable of withstanding temperatures above 450 °C.

DOI: 10.1201/9781003247005-10

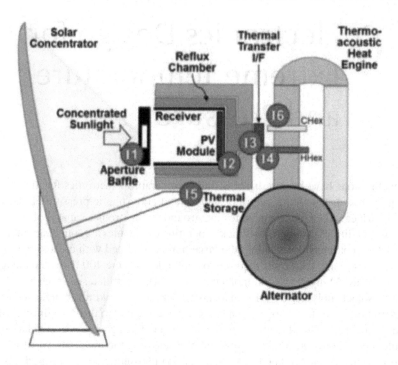

FIGURE 10.1 Solar concentrator system architecture and thermal interfaces.

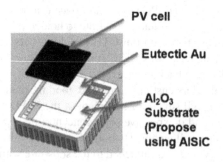

FIGURE 10.2 High temperature (+300 °C) packaging stack-up for PV cell.

REFERENCES

1. Grandidier, J., B.J. Nesmith, T.J. Hendricks, and J. Cepeda-Rizo, "Full spectrum hybrid photovoltaics and thermal engine utilizing high concentration solar energy," *33rd European Photovoltaic Solar Energy Conference and Exhibition*, EU PVSEC 2016, June 20–24, 2016, Munich.
2. Cepeda-Rizo, J., and T. Hendricks, "Naphthalene boiling and condensation heat transfer performance enabling hybrid solar power systems," *ASME 2016 Heat Transfer, Fluids Engineering, & Nanochannels, Microchannels, and Minichannels Conferences*, Washington, DC, July 10–14, 2016.

11 Characterization and Modeling of PWB Warpage and Its Effect on LGA Separable Interconnects

INTRODUCTION

The land grid array (LGA) connector, as shown in Figure 11.1, uses compression to achieve an interconnect between the module and the printed wiring board (PWB). Essentially, the LGA connector is a replacement for soldered connections and allows for the frequent removal and replacement of the module on the daughter card. These connectors also provide electrical advantages due to their low profile and compressibility, and their simplicity offers cost-saving advantages [1]. There are many LGA connectors on the market with pins ranging from tellurium copper c-clip style to elastomeric buttons.

A study was performed to investigate PWB-warpage effect maintaining an adequate force distribution across all the contacts of an LGA connector and to test a fastening and supporting hardware's ability to distribute the force sufficiently—and consistently—by manipulating the board contour.

Flatness of the sandwiched elements (i.e., module and PWB) is critical to the ability of the LGA to achieve adequate contact. The substrate on the module is ceramic and maintains a reliable amount of flatness. However, the PWB is a laminated epoxy matrix composite notorious for warping and bowing. Of the factors that contribute to the overall unevenness of the system—PWB warpage, cold plate/module bow, and plating unevenness of the module array pads on the PWB, PWB warpage tends to be the leading contributor and the subject of this investigation.

The intent of this work is to model the effects of warpage on a PCB due to placement of the connector with predetermined fastening hardware and to determine if this warpage is acceptable to the performance of the connector. A conventional analytic approach was used to model the fastening hardware, and experimental data was collected to verify the models. After analysis of the results, classical laminate theory was then used to better address the warpage that was observed through the experiment.

DOI: 10.1201/9781003247005-11

125

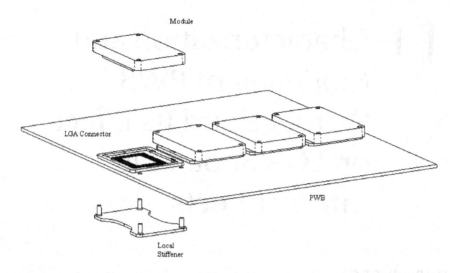

FIGURE 11.1 Electronic module assembly on a PWB. An individual stack-up is composed of module, LGA connector, PWB, and local stiffener plate.

MECHANICAL ASPECTS OF AN LGA

The pins of an LGA behave like springs with a certain limited range of travel and stiffness (Figure 11.1). The LGA pins used in this chapter have a total compliance of 254 microns (.010 inch), with an estimated linear stiffness of 0.4 gram-force/micron (10 gram-force/.001inch). The connector has a manufacturer's recommended operation range of 40–80 gram-force, with a not-to-exceed maximum load of 120 gram-force. In order to achieve the minimum operation load for sufficient contact, a preload deflection of 100 microns (.004 inch) is required, which reduces the amount of total pin travel from 250 microns (.010 inch) to 152 microns (.006 inch). It is anticipated that variation in the module assembly components (i.e., stack-up tolerances, fastener tolerances, etc.), not including PWB, will require a maximum of ±45 microns (±.0018 inch) of pin travel 91 microns (.0036 inch) unilaterally, leaving approximately 61 microns (.0024 inch) of total pin travel for variation in the PWB (i.e., warpage allowance). This variation amounts to 2.3 microns per mm (.0023 inch per inch) maximum PWB curvature (1.2% maximum IPC [Institute of Interconnecting and Packaging Electronic Circuits] twist for a 510-centimeter (20-inch) board) for the connector operation range. In other words, the PWB could take on a bow equivalent to 1.2% as measured by the IPC method and still be able to operate within the 40–80 gram-force per pin working range. Another way to look at it is that the board requires 60 microns (.0024 inch) of local "planarity" or flatness as measured off LGA pads on the PWB.

FASTENING HARDWARE

The module/LGA assembly is attached to the PWB by the use of springs and stand-offs designed to transmit 60 gram-force per pin on the LGA. The local stiffener

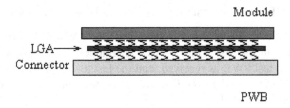

FIGURE 11.2 Module/LGA/PWB system. LGA pins behave like small springs sandwiched between module and PWB components.

FIGURE 11.3 Impression of an LGA using Fuji pressure sensitive film and converted to stress marks by densimeter scanner. Stress marks are then interpreted by use of a color map.

plate is a low-profile flat spring designed to minimize the local warping of the PWB caused by the total clamping force of the assembly.

METHOD OF INVESTIGATION

PRESSURE SENSITIVE FILM

Fuji pressure sensitive film was used to measure an imprint of the force distribution across all the LGA pins as well as pressure distribution of the stiffener plate to the PWB. The film employs the use of micro-size liquid-filled vacuoles that release ink upon compression. The density of the ink released is proportional to the amount of pressure applied. A color swatch in combination with charts that take temperature and humidity into consideration are used to convert color intensity into an actual pressure value. The film can also be analyzed by a scanner equipped with a densimeter that converts the impression into stress marks on the film (Figure 11.3).

PWB Warpage/Flatness Measurements Before Loading of Modules

Three different methods were used to determine the preloaded warpage of the test PWB depending on the test. The first two techniques listed next were performed on all bare board test PWB as received, and the last technique mentioned was done on only two PWBs to see the overall warping and was intended to characterize the entire lot of boards received.

Method I: IPC TM-650 (Interconnect and Packaging Committee-Test Method)

This is performed by internal inspection department on a sample-lot basis and is the only warpage assessment technique employed to determine adherence to [2]. The technique involves laying the panel on a flat granite datum surface and using a pin gauge at designated areas to measure the panel profile. It is a quick way to obtain a warpage value that can be used to accept or reject material base from a specification. The dial indicator has an accuracy of ±.0001 inch.

Method II: LGA Area Grid Profile

The LGA matrix grid profile was accomplished by drawing a 10.67 × 15.88 mm (0.420 × 0.625 in.) grid across the PWB in and around the LGA areas, and measuring Z-axis values as a function of X and Y, approximately 500 data points were taken. This data is then inputted into a surface-modeling program where cubic splines are created and then merged together to create a topographical surface profile. Because actual measurements could not be taken directly off the PWB underneath the module/LGA, this technique essentially estimated the curvature at this area by best fitting the cubic spline surface through adjacent points.

Method III: Cooper Thin-Film Load Sensors

The connector was removed and in its place six Cooper sensors (Cooper Instruments model ELF 4200—Flexiforce sensor, medium 25–150 lb range) were attached to the PCB (Figure 11.1). The sensors are 9.5 mm (0.375 in.) in diameter, 0.23 mm (.009 in.) thick, and use a conductive ink that when compressed records a change in electrical resistance; this resistance is converted into a force value by means of calibration with known applied loads. A compliant silicone pad was used to distribute the load evenly over the sensing surface of the sensor. By means of shimming, the overall thickness of the sensor with pads was made equal to that of the LGA connector.

Ten measurements were taken at one module site on a test board. The force reading was converted into a local pressure for each sensor and a force per pin was calculated. Each sensor was positioned to encircle 38 pins.

WARPAGE PREDICTION USING CLASSICAL LAMINATE THEORY

To understand the nature of the board warpage better, board warpage was investigated using classical laminate theory. The test board is composed of 24 layers that are 0 degree and 90 degree, thus making the laminate both balanced and symmetrical. Because there is no thermal expansion loading nor are there axial forces, the constitutive equations reduce down to those associated with laminate transverse bending [3]:

$$
\begin{bmatrix} M_{xx} \\ M_{yy} \\ M_{xy} \end{bmatrix} = \begin{bmatrix} D_{11} & D_{12} & D_{16} \\ D_{12} & D_{22} & D_{26} \\ D_{16} & D_{26} & D_{66} \end{bmatrix} \begin{bmatrix} \dfrac{\partial^2 w}{\partial x^2} \\ \dfrac{\partial^2 w}{\partial y^2} \\ 2\dfrac{\partial^2 w}{\partial x \partial y} \end{bmatrix}
\tag{11.1}
$$

where M_{xx}, M_{yy}, and M_{xy} are the moment resultants, D_{ij} are the bending stiffnesses, and w_o is the transverse displacement of the midplane of the laminate. Multiplying the bending stiffness matrix with the transverse displacement vector yields:

$$
\begin{bmatrix} M_{xx} \\ M_{yy} \\ M_{xy} \end{bmatrix} = \begin{array}{l} D_{11}\dfrac{\partial^2 w}{\partial x^2} + D_{12}\dfrac{\partial^2 w}{\partial y^2} + 2D_{16}\dfrac{\partial^2 w}{\partial x \partial y} \\[2mm] D_{12}\dfrac{\partial^2 w}{\partial x^2} + D_{22}\dfrac{\partial^2 w}{\partial y^2} + 2D_{26}\dfrac{\partial^2 w}{\partial x \partial y} \\[2mm] D_{16}\dfrac{\partial^2 w}{\partial x^2} + D_{26}\dfrac{\partial^2 w}{\partial y^2} + 2D_{66}\dfrac{\partial^2 w}{\partial x \partial y} \end{array}
\tag{11.2}
$$

For steady-state conditions, neglecting thermal effects, and assuming no in-plane forces applied to the laminate, the equation of equilibrium of the laminate [4] is:

$$
\frac{\partial^2 M_{xx}}{\partial x^2} + 2\frac{\partial^2 M_{xy}}{\partial x \partial y} + \frac{\partial^2 M_{yy}}{\partial y^2} + q = 0
\tag{11.3}
$$

where q is the distributed transverse load on the plate.

Substitution of Equation (11.2) into (11.3) gives the governing equation for steady-state conditions with no in-plane applied forces as [5, 6]:

$$
D_{11}\frac{\partial^4 w_o}{\partial x^4} + 2(D_{12} + 2D_{66})\frac{\partial^4 w_o}{\partial x^2 \partial y^2} + D_{22}\frac{\partial^4 w_o}{\partial y^2} + q = 0
\tag{11.4}
$$

Try Navier solution [5, 7] for a simply supported plate on all edges with boundary conditions:

$$
w(x,y) = \sum_{m=1}^{\infty} \sum_{n=1}^{\infty} w_{mn} Sin\frac{m\pi x}{a} Sin\frac{n\pi y}{b}
\tag{11.5}
$$

$$w(a,0) = 0 = w,_{xx}(a,0)$$
$$w(0,b) = 0 = w,_{yy}(0,b)$$
$$w(a,y) = 0 = w,_{xx}(a,y)$$
$$w(x,b) = 0 = w,_{yy}(x,b)$$

(11.6)

It can be readily seen that the boundary conditions are satisfied.
The load q_{mn} is also approximated by a double Fourier series:

$$q(x,y) = \sum_{m=1}^{\infty} \sum_{n=1}^{\infty} q_{mn} Sin \frac{m\pi x}{a} Sin \frac{n\pi y}{b}$$

(11.7)

where

$$q_{mn} = \frac{4}{ab} \int_0^b \int_0^a q(x,y) Sin \frac{m\pi x}{a} Sin \frac{n\pi y}{b} dxdy$$

(11.8)

$$w_{mn} = \frac{1}{\pi^4} q_{mn} \left[D_{11} \left(\frac{m}{a} \right)^4 + 2(D_{12} + 2D_{66}) \left(\frac{mn}{ab} \right)^2 + D_{22} \left(\frac{n}{b} \right)^4 \right]^{-1}$$

The transverse deflection w_{mn} is obtained by substituting (11.5) into the governing equation as well as (11.7) and (11.8).

Based on the balanced and symmetrical properties of the laminate, a simplified first order approximation of isotropic stiffness properties was assumed.

$$w_{mn} = \frac{1}{\pi^4} q_{mn} D \left[\left(\frac{m}{a} \right)^2 + \left(\frac{n}{b} \right)^2 \right]^{-2}$$

where
E = 20.7 GPa;   Young's modulus
v = 0.16   Poisson's ratio
h = 3.18 mm;   Board thickness

$$D = \frac{Eh^3}{12(1-v^2)} \qquad \text{flexural rigidity of the plate}$$

To make further comparison consider the Levy solution [5, 7]:

$$w = \sum_{m=1}^{\infty} Y_m \sin \frac{m\pi x}{a}$$

(11.9)

where Y_m is a function of y only [5, 7]. It is assumed that the sides $x = 0$ and $x = a$ are simply supported. Hence, each term of series (11.9) satisfies the boundary conditions $w = 0$ and $\dfrac{\partial^2 w}{\partial x^2} = 0$ at these two sides. It remains to determine Y_m in such a form as to satisfy the boundary conditions on the sides $y = \pm b / 2$ and also the equation of deflection surface $\dfrac{\partial^4 w}{\partial x^4} + 2\dfrac{\partial^4 w}{\partial x^2 \partial y^2} + \dfrac{\partial^4 w}{\partial y^4} = \dfrac{q}{D}$. The fact that the Levy solution only requires two simply supported edges and the other two edges could be defined arbitrarily makes it more versatile than the Navier approximation.

RESULTS

IPC Measurements

The test board was measured to have 0.2% of initial bow per IPC-TM-60, which translates to a local 13 microns (.0005 inch) of out-of-plane deflection per LGA on the PWB. This is almost five times flatter than what is required. This amount of flatness was expected due to the balanced nature of the PWB laminate of this test board.

LGA Area Grid Profile

Two measurements were taken, one before assembly, and the other after, using the grid profile technique described previously. For visual clarity, the spline surfaces were created using z-values that were multiplied by 50 (see Figure 11.4). The modules cause the board to take on a local warpage that was repeated at every module location. The center of the LGA area on the board was bowed towards the module (tangent to the module), while it was noticed that the board area at the leading edge of the connector bowed away from the module.

The local nature of the warpage, as shown in Figure 11.5, is seen to be like a reverse "saddle" type of bow. The out-of-plane displacement at this area is −76 microns (−.003 inch), which is larger than the maximum allowed out-of-plane board displacement of 61 microns (.0024 inch).

Cooper Sensor and Fuji Pressure Film Measurements

The areas on the PWB test board where the sensors were attached are shown in Figure 11.6. Table 11.1 shows the results of the Cooper sensor measurements and Table 11.2 shows the results of the Fuji film measurements. The lowest readings are

FIGURE 11.4 Surface plot of PWB cubic splines as measured with no modules attached (undeflected). Magnified 50x in transverse (out-of-plane) direction for clarity. Darker framed areas are where LGA attaches.

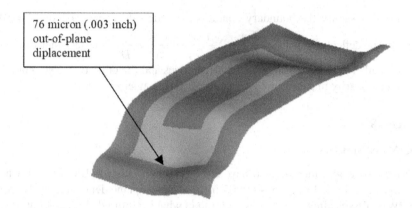

FIGURE 11.5 Close-up of Section 1 of PWB showing a reverse "saddle" shape warpage, where the center of the array is bowing outward (convex) and the leading edge is bowing inward (convex). Magnified z-direction by 50x.

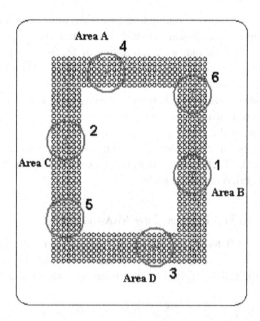

FIGURE 11.6 Cooper sensor placements onto PWB test board. Circles indicate where the six sensors resided.

Area A and Area D, which are consistent with the areas of greatest measured negative out-of-plane displacement, or low spots, of the PWB.

NAVIER APPROXIMATION

A simple program was written in MATLAB 6.0 code to solve Equation (11.4). Successive iterations were calculated until the result converged to within four significant figures.

TABLE 11.1
Results of the Cooper Sensor Values in Gram-Force per Pin

Sensor #	1 (Area B)	2 (Area C)	3 (Area D)	4 (Area A)	5	6
run1	66.7	87.8	37.3	27.0	75.1	76.3
run2	69.1	83.4	40.7	31.4	81.4	78.8
run3	70.9	85.0	39.7	30.6	77.7	82.4
run4	69.9	90.4	41.3	26.6	73.5	76.9
run5	70.9	86.2	40.7	31.6	75.1	79.8
run6	69.1	85.0	42.7	29.6	80.0	84.0
run7	74.9	89.8	41.3	27.0	75.3	81.2
run8	71.7	93.7	42.7	27.6	69.5	81.6
run9	70.9	88.8	43.1	28.6	75.5	81.6
run10	74.1	92.4	39.3	26.0	74.1	83.0

TABLE 11.2
Results of Fuji Film in Gram-Force per Pin

Run	Area A	Area B	Area C	Area D
1	36.5	82.1	76.6	84.3
2	42.2	79.8	81.3	50.1
3	40.0	79.9	82.3	49.4

The three-dimensional plot of the Navier approximation is shown in Figure 11.7. The loads transmitted by the module (through the LGA) and stiffener plate to the PWB were treated as superimposed loads and inputted into Equation (11.8) [8]. The loading footprint dimension of the stiffener plate to the PWB was obtained with Fuji pressure film. The result was a reverse "saddle" bending of the plate similar to that measured using the test board. The area of low force, Area "A," had an out-of-plane deflection of −53 microns (−.0021 inch), which though lying within the −61-micron (−.0024-inch) maximum allowable working range of the connector, is close to this maximum. The plot showed the center of the LGA pin area, Areas B and C, to have a positive out-of-plane deflection of 27 microns (.0011 inch), which lie well within the acceptable flatness range. The actual center of the PWB underneath the module measured 64 microns (.0025 inch) of out-of-plane displacement, but because the LGA has no contacts in this center area, there is no concern of inadequate force distribution.

LEVY APPROXIMATION

The alternative solution, the form shown in Equation (11.9), used the MATLAB 6.0 code to solve Equation (11.4). Successive iterations were calculated until the result

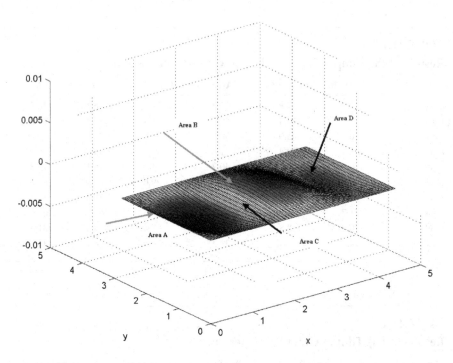

FIGURE 11.7 PWB warpage by Navier's Fourier approximation predicting the reverse saddle warpage of the board at the LGA area. All dimensions shown are in inches.

converged to within four significant figures. The three-dimensional plot of the Levy approximation is shown in Figure 11.8. The benefit of the Levy approximation is that it treats the sides adjacent to Area A and Area D as free edges, which is deemed more closely representative of the true conditions of the PWB. The approximation was obtained by superposition of the out-of-plane deflection of a partially loaded simply supported rectangular plate with the deflection of a homogeneous solution with edge loads devised to cancel out the resultant shear forces at the edges adjacent to Area A and Area D. This resulted in a solution that approximated the out-of-plane deflection of rectangular plate with two sides simply supported and two sides free. The result was a reverse "saddle" bending of the plate similar to that measured using the test board and seen on the Navier approximation. The area of low force, Area "A," had an out-of-plane deflection of −74 microns (−.0029 inch), which was larger than that approximated by the Navier solution. The plot showed the center of the LGA pin area, Areas B and C, to have a positive out-of-plane deflection of 25 microns (.0010 inch), which lies well within the acceptable flatness range.

COMPARISON OF PREDICTED (NAVIER SOLUTION) AND EXPERIMENTAL VALUES

As shown in Table 11.3, the predicted displacement of −53 microns (−.0021 inch) in Area A was lower than the as-measured displacement of −76 microns (−.003 inch). Area B showed a predicted displacement of 27 microns (.0011 inch), which also

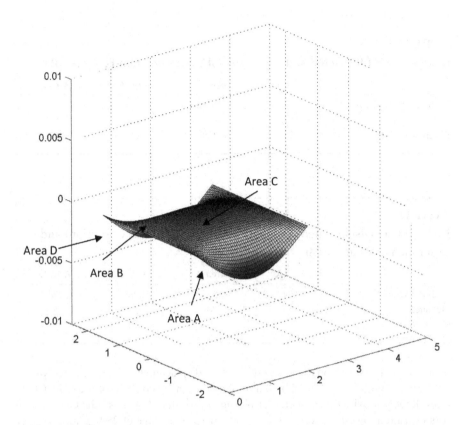

FIGURE 11.8 PWB warpage by Levy's approximation predicting the reverse saddle warpage of the board at the LGA area. All dimensions shown are in inches.

TABLE 11.3

Out-of-Plane Displacement Comparison of Predicted (Navier Solution) and Experimental Values in Microns

	Area A	Area B	Area C	Area D
Predicted Values (Navier)	−53	27	27	−38
Measured	−76	20	23	−50
% of Errors	30.3	35	21.7	24

resulted in significant error compared to the measured value. The general reverse "saddle" bending contour of the PWB locally to one electronic module was apparent in both the measured data (Figure 11.5) and the predicted data (Figure 11.7).

Table 11.4 shows the predicted versus experimental values for the force per pin of the LGA. The assembly hardware was designed to nominally achieve 60 gram-force per pin of force; board warpage will cause the pin forces to vary above and below

TABLE 11.4

Comparison of Predicted and Experimental Values in Gram-Force per Pin

	Area A	Area B	Area C	Area D
Predicted Values (Navier)	38.1	71.5	71.5	44.4
Experiment (Fuji)	31.6	64.5	64.0	49.0
Experiment (Cooper)	28.6	70.8	88.3	40.9

TABLE 11.5

Out-of-Plane Displacement Comparison of Predicted (Levy Solution) and Experimental Values in Microns

	Area A	Area B	Area C	Area D
Predicted Values (Levy)	−74	25	25	−74
Measured	−76	20	23	−50
% of Errors	2.6	20	10	48

the nominal depending on the local out-of-plane displacements. The predicted values for force per pin were obtained by multiplying the spring constant of the LGA (10.41 gram-force/pin) with the predicted displacements of the LGA pins. The Cooper sensors measured lowest in Area A (Sensor 4) where an average of 28.6 gram-force per pin was apparent; this is outside of the working range of 40–80 gram-force for the LGA connector. In comparison, Fuji film analysis was taken in Area A and resulted in an average value of 31.6 gram-force. Both sensor methods pointed to a low force per pin reading in Area A, which implied a negative out-of-plane deflection of the PWB. An anomalous reading of the Cooper sensor (88.3 gram-force per pin) was observed at Area C, which was much larger than the predicted value of 71.5 gram-force and the Fuji film measurement of 64.0 gram-force. Multiple Cooper sensors were substituted in that area and all read comparable values. Also, as mentioned previously, a compliant silicone shim was used in place of the LGA connector to distribute the force over the sensor area, and though the material was selected to closely match the stiffness of the LGA pin, it may have not fully represented the characteristics of the pins.

COMPARISON OF PREDICTED (LEVY SOLUTION) AND EXPERIMENTAL VALUES

As shown in Table 11.5, the predicted displacement of −74 micron (−.0029 inch) in Area A was lower than the as-measured displacement of −76 microns (−.003 inch). Area B showed a predicted displacement of 25 microns (.0010 inch) compared to the measured value of 20 microns (.0008 inch). The general reverse "saddle" bending contour of the PWB locally to one electronic module was apparent in both the

TABLE 11.6
Comparison of Predicted (Levy Solution) and Experimental Values in Gram-Force per Pin

	Area A	Area B	Area C	Area D
Predicted Values (Levy)	29.8	70.4	70.4	29.8
Experiment (Cooper)	28.6	70.8	88.3	40.9

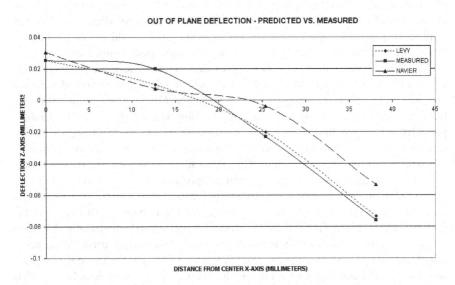

FIGURE 11.9 Comparison of both Navier and Levy predicted approximations vs. the measured values. The horizontal axis represents the distance from the center of the plate along the *x*-axis.

measured data (Figure 11.5) and in the predicted data (Figure 11.8). The deflection of the free edges A and D were also apparent in Figure 11.8 and provided a better approximation than Navier did.

Table 11.6 shows the predicted versus experimental values for the force per pin of the LGA. The free-edge conditions near A and B resulted in closer approximation than did Navier to the actual deflection, where an error of 2.6% was calculated. Due to the complexity of the Levy approximation, symmetry was assumed about the *x*-axis, which resulted in the deflections at A and B being equal to those at C and D, respectively. As a result, the largest measured deflection, Area A, was approximated with good accuracy, while the deflection at D was less well represented.

The overall comparison between both Navier and Levy prediction versus experimentally measured data are illustrated in Figure 11.9. The Levy approach offered better estimation of the PWB contour compared to Navier, especially near the free edges of the plate (Area A). Areas B and C were also better characterized by Levy than by Navier. To obtain the improved prediction, numerical solutions for the aniso-tropic laminated plate will be necessary.

CONCLUSION

The results of the experimental measurements and of the laminate model analysis indicated a need for additional stiffness in Areas A and D of the PWB by the using either a rib-type stiffener or a change of design of the backside module stiffener. The loading conditions on the LGA cause a "saddle"-shaped warpage local to the PWB, resulting in negative out-of-plane deflection and thus low force readings in those areas. From the first-order simplified analytical solution, the predicted positive out-of-plane displacement observed at the center of the LGA would not jeopardize the functionality of the connector because of the absence of pins in that area.

The classical laminate theory model provided a reasonable approximation of the actual overall contour of the PWB local to the module when compared with the experimental surface measurements. The predicted Navier model underestimated the deflection by 30%, as apparent in Area A. However, the predicted Levy model predicted a much smaller error of 2.6%. Both the Navier and the Levy approximation models consisted of simple MATLAB programming that used minimal computa-tional resources to provide an output. The Navier model consisted of 20 lines of code, while the Levy model was much longer and complicated and converged rather slowly. For example, the Navier model required only four terms to achieve four significant figures of convergence, while the Levy model required about 14. To overcome the slow convergence, simplification of the model was performed where applicable, such as the assumption of symmetry about the x-axis. It was also apparent that concen-trated loads tended to destabilize the Levy model especially at the free edges, so care was taken to characterize the loading conditions properly. Nonetheless, both models provide an elegant approach to predicting out-of-plane plate displacements for well-known and measurable loading conditions.

The Fuji pressure film and Cooper pressure sensors served well in identifying these loading conditions, as well as in identifying the areas of low force on the LGA connector. The advantage the Fuji film had over the Cooper sensors is that it is able to take a reading in situ of the LGA loading conditions. The Cooper sensor provided an indirect measurement of the LGA force per pin, in that a compliant shimming mate-rial is used in place of the LGA connector. One particular location of the LGA array on the PWB, Area C, measured uncharacteristically high values using the Cooper sensors, which was not reflected in the Fuji film measurements.

ACKNOWLEDGEMENTS

The author would like to thank Teradyne for providing facilities to this investigation. Also, special thanks to Tarzen Kwok, Herman Heinrich, and Steven Allameh for their help designing the test vehicles, and to Marlon Jimenez for help with the Cooper sensors.

REFERENCES

1. Pecht, M., et al., "Assessing the operation reliability of land grid array elastomer sockets," *IEEE Transactions on Component and Packaging Technologies*, vol. 23, no. 1, March 2000, p. 171.
2. *IPC TM—650 Test Procedure for Measuring PCB Bow and Twist* © 2001 IPC—Association Connecting Electronics Industries 2215 Sanders Road, Northbrook, IL 60062–6135.
3. Polsky, Y., and C. Ume, "Comparison of first-order shear and plane strain assumptions in warpage prediction of simply supported printed wiring boards," *ASME Journal of Electronic Packaging*, vol. 123, March 2001, pp. 1–5.
4. Suhir, E., *Structural Analysis in Microelectronic and Fiber-Optic Systems*, Vol. 1, AT&T Bell Laboratories, Van Nostrand Reinhold, New York, 1991, p. 260.
5. Timoshenko, S., and S. Woinowsky-Krieger, *Theory of Plates and Shells*, McGraw-Hill, New York, June 1959, p. 113.
6. Timoshenko, S., and Goodier, J.N., *Theory of Elasticity,* 3rd Ed., McGraw-Hill Book Co., Inc., New York, 1970.
7. McFarland, D., B. Smith, and W. Bernhart, *Analysis of Plates*, Spartan Books, New York, 1972, p. 6
8. Esfandiari, R., *Applied Mathematics for Engineers,* 2nd Ed., McGraw-Hill, New York, 1996, p. 369.

APPENDIX A

Levy Solution Derivations for the Biharmonic Equation for Rectangular Plate With Opposite Sides Simply Supported, and Two Sides Free, With Partial Loading—Chapter 1

For localized loading, Timoshenko [5] gives:

$$
w_1 = \frac{4qa^4}{D\pi^5} \sum_{1,3,5}^{\infty} \frac{(-1)^{\frac{m-1}{2}}}{m^5} \sin\frac{m\pi a}{2a}
$$

$$
\left\{
\begin{array}{l}
1 - \dfrac{\cosh\dfrac{m\pi y}{a}}{\cosh\alpha_m}\left[\cosh(\alpha_m - 2\gamma_m) + \alpha_m \dfrac{\sinh 2\gamma_m}{2\cosh\alpha_m}\right] + \\[4mm]
\dfrac{\cosh(\alpha_m - 2\gamma_m)}{2\cosh\alpha_m}\dfrac{m\pi y}{a}\sinh\dfrac{m\pi y}{a}
\end{array}
\right\} \sin\frac{m\pi x}{a}
\qquad\text{(A.1)}
$$

where w_1 represents partial loading for a simply-supported plate.

Find w_2 for deflection caused by edge shear in terms of E_m. w_2 represents out-of-plane deflection of a simply supported plate with edge shear on two opposing sides. $z = w_1 + w_2$ is the superposition that will result in partial loading of a plate with two opposing simply supported sides and two opposing free sides.

$$
\left[\frac{\partial^3 w_2}{\partial y^3} + (2-v)\frac{\partial^3 w_2}{\partial\partial^2 y}\right]_{y=\frac{b}{2}} = \sum_{m=1,3,5}^{\infty} E_m \sin\frac{m\pi x}{b}
\qquad\text{(A.2)}
$$

$$
w_2 = \sum_{m=1,3,5}^{\infty}\left(B_m \cosh\frac{m\pi y}{a} + C_m \frac{m\pi y}{a}\sinh\frac{m\pi y}{a}\right)\sin\frac{m\pi x}{b}
\qquad\text{(A.3)}
$$

$$
\left(\frac{\partial^2 w_2}{\partial^2 y^2} + v\frac{\partial^2 w_2}{\partial^2 y^2}\right)_{y=\frac{b}{2}} = 0
\qquad\text{(A.4)}
$$

$$
\frac{\partial w}{\partial y} = \sum_{m=1,3,5}^{\infty}\left(B_m\frac{m\pi}{a}\sinh\frac{m\pi y}{a} + C_m\left[\frac{m\pi y}{a}\left(\frac{m\pi}{a}\right)\cosh\frac{m\pi y}{a} + \left(\frac{m\pi}{a}\right)\sinh\frac{m\pi y}{a}\right]\right)
$$

$$
\sin\frac{m\pi x}{b}
$$

$$
\qquad\text{(A.5)}
$$

$$\frac{\partial^2 w}{\partial^2 y} = \sum_{m=1,3,5}^{\infty} \left(B_m \left(\frac{m\pi}{a}\right)^2 \cosh\frac{m\pi y}{a} + C_m \left[\frac{\left(\frac{m\pi}{a}\right)^2 \frac{m\pi y}{a}\sinh\frac{m\pi y}{a} + \left(\frac{m\pi}{a}\right)\cosh\frac{m\pi y}{a} +}{\left(\frac{m\pi}{a}\right)^2 \cosh\frac{m\pi y}{a}}\right]\right)$$
$$\sin\frac{m\pi x}{b} \tag{A.6}$$

$$v\frac{\partial^2 w}{\partial^2 x} = \sum_{m=1,3,5}^{\infty} \left(B_m \cosh\frac{m\pi y}{a} + C_m\left[\frac{m\pi y}{a}\sinh\frac{m\pi y}{a}\right]\right)\left(-v\left(\frac{m\pi}{a}\right)^2\right)\sin\frac{m\pi x}{a} \tag{A.7}$$

$$\sum_{m=1,3,5}^{\infty} \left(B_m \left(\frac{m\pi}{a}\right)^2 \cosh\frac{m\pi y}{a} + C_m \left[\frac{\left(\frac{m\pi}{a}\right)^2 \frac{m\pi y}{a}\sinh\frac{m\pi y}{a} + \left(\frac{m\pi}{a}\right)\cosh\frac{m\pi y}{a} + \left(\frac{m\pi}{a}\right)^2}{\cosh\frac{m\pi y}{a}}\right]\right)$$
$$\sin\frac{m\pi x}{b} +$$
$$\sum_{m=1,3,5}^{\infty} \left(B_m \cosh\frac{m\pi y}{a} + C_m\left[\frac{m\pi y}{a}\sinh\frac{m\pi y}{a}\right]\right)\left(-v\left(\frac{m\pi}{a}\right)^2\right)\sin\frac{m\pi x}{a} = 0 \tag{A.8}$$

$$B_m = \frac{-C_m\left(\frac{m\pi y}{a}\sinh\frac{m\pi y}{a} + 2\cosh\frac{m\pi y}{a} - v\frac{m\pi y}{a}\sinh\frac{m\pi y}{a}\right)}{(1-v)\cosh\frac{m\pi y}{a}} \tag{A.9}$$

$$B_m = \frac{-C_m\left(\alpha_m \sinh\alpha_m + 2\cosh\alpha_m - v\alpha_m \sinh\alpha_m\right)}{(1-v)\cosh\alpha_m} = -C_m\Phi \tag{A.10}$$

$$w_2 = \sum_{m=1,3,5}^{\infty} C_m\left(\frac{m\pi y}{a}\sinh\frac{m\pi y}{a} - \Phi\cosh\frac{m\pi y}{a}\right)\sin\frac{m\pi x}{a} \tag{A.11}$$

$$\frac{\partial^2 w_2}{\partial^2 y} = \sum_{m=1,3,5}^{\infty} C_m \left(\begin{pmatrix} \frac{m\pi}{a}\left(\frac{m\pi y}{a}\right)\cosh\frac{m\pi y}{a}+\left(\frac{m\pi y}{a}\right)\sinh\frac{m\pi y}{a} \\ -\Phi\left(\frac{m\pi}{a}\right)\sinh\frac{m\pi y}{a} \end{pmatrix} \right) \sin\frac{m\pi x}{a}$$ (A.12)

$$\frac{\partial^2 w_2}{\partial^2 y} = \sum_{m=1,3,5}^{\infty} C_m \left(\left(\frac{m\pi}{a}\right)^2 \begin{pmatrix} \frac{m\pi y}{a}\sinh\frac{m\pi y}{a}+\cosh\frac{m\pi y}{a}+\left(\frac{m\pi}{a}\right)^2 \\ \cosh\frac{m\pi y}{a}-\Phi\left(\frac{m\pi}{a}\right)^2\cosh\frac{m\pi y}{a} \end{pmatrix} \right) \sin\frac{m\pi x}{a}$$ (A.13)

$$\frac{\partial^2}{\partial x^2}\left(\frac{\partial w_2}{\partial y}\right) = \sum_{m=1,3,5}^{\infty} -C_m\left(\frac{m\pi}{a}\right)^3\left[\frac{m\pi y}{a}\cosh\frac{m\pi y}{a}+(1-\Phi)\sinh\frac{m\pi y}{a}\right]\sin\frac{m\pi x}{a}$$ (A.14)

$$\frac{\partial^3 w_2}{\partial^3 y} = \sum_{m=1,3,5}^{\infty} \left(C_m\left(\left(\frac{m\pi}{a}\right)^3\left(\frac{m\pi y}{a}\cosh\frac{m\pi y}{a}+(3-\Phi)\sinh\frac{m\pi y}{a}\right)\right) \right)\sin\frac{m\pi x}{a}$$ (A.15)

$$\left[\frac{\partial^3 w_2}{\partial y^3}+(2-v)\frac{\partial^3 w_2}{\partial x^2 \partial y}\right]_{y=\frac{b}{2}} = \sum_{m=1,3,5}^{\infty}\left(E_m\sin\frac{m\pi x}{a}\right)$$ (A.16)

$$C_m = \frac{E_m}{\Psi}, \alpha_m = \frac{m\pi b}{2a}$$ (A.17)

where $\Psi = \left(\frac{m\pi}{a}\right)^3\begin{bmatrix}\alpha_m\cosh\alpha_m+(3-\Phi)\sinh\alpha_m-\\(2-v)(\alpha_m\cosh\alpha_m+(1-\phi)\sinh\alpha_m)\end{bmatrix} = \text{constant}$

$$w_2 = \sum_{m=1,3,5}^{\infty}\frac{E_m}{\Psi}\left[\frac{m\pi y}{a}\sinh\frac{m\pi y}{a}-\Phi\cosh\frac{m\pi y}{a}\right]\sin\frac{m\pi x}{a}$$ (A.18)

Back to localized loading (Timoshenko [5])

$$w_1 = \frac{4qa^4}{D\pi^5} \sum_{1,3,5}^{\infty} \frac{(-1)^{\frac{m-1}{2}}}{m^5} \mathrm{Sin}\frac{m\pi a}{2a} \left\{ 1 - \frac{\cosh\dfrac{m\pi y}{a}}{\cosh\alpha_m} \left[\begin{array}{l} \cosh\left(\alpha_m - 2\gamma_m\right) + \\ \alpha_m \dfrac{\sinh 2\gamma_m}{2\cosh\alpha_m} \end{array} \right] + \frac{\cosh\left(\alpha_m - 2\gamma_m\right)}{2\cosh\alpha_m} \frac{m\pi y}{a}\sinh\frac{m\pi y}{a} \right\} \sin\frac{m\pi x}{a}$$

(A.1)

$$w_1 = \left\{ 1 - J\cosh\frac{m\pi y}{a} + Q\frac{m\pi y}{a}\sinh\frac{m\pi y}{a} \right\} \frac{m\pi x}{a}$$

(A.19)

where

$$K = \frac{4qa^4}{D\pi^2 5}\frac{(-1)^{\frac{m-1}{2}}}{m^5}\sin\left(\frac{m\pi a1}{2a}\right),$$

$$J = \frac{\left[\cosh\left(\alpha_m - 2\gamma_m\right) + \gamma_m\sinh\left(\alpha_m - 2\gamma_m\right) + \dfrac{\alpha_m\sinh 2\gamma_m}{2\cosh\alpha_m}\right]}{\cosh\alpha_m}$$

(A.20)

$$Q = \frac{\cosh\left(\alpha_m - 2\lambda_m\right)}{2\cosh\alpha_m}, \gamma_m = \frac{m\pi b_1}{4a}$$

(A.21)

$$\frac{\partial w_1}{\partial y} = \sum_{m=1,3,5}^{\infty} K \left[\frac{J\left(\dfrac{m\pi}{a}\right)\sinh\dfrac{m\pi y}{a} + Q\left(\dfrac{m\pi y}{a}\right)\left(\dfrac{m\pi}{a}\right)}{\cosh\left(\dfrac{m\pi y}{a}\right) + \left(\dfrac{m\pi}{a}\right)\sinh\dfrac{m\pi y}{a}} \right] \sin\frac{m\pi x}{a}$$

(A.22)

$$\frac{\partial^2 w_1}{\partial y^2} = \sum_{m=1,3,5}^{\infty} K \left[\frac{J\left(\dfrac{m\pi}{a}\right)^2\cosh\dfrac{m\pi y}{a} + Q\left(\dfrac{m\pi y}{a}\right)\left(\dfrac{m\pi}{a}\right)^2}{\sinh\left(\dfrac{m\pi y}{a}\right) + \left(\dfrac{m\pi}{a}\right)^2\cosh\left(\dfrac{m\pi y}{a}\right)} + \left(\dfrac{m\pi}{a}\right)^2\cosh\left(\dfrac{m\pi y}{a}\right) \right] \sin\frac{m\pi x}{a}$$

(A.23)

$$
\frac{\partial^3 w_1}{\partial y^3} = \sum_{m=1,3,5}^{\infty} K\left(\frac{m\pi}{a}\right)^3 \left[\begin{array}{l} J \sinh\dfrac{m\pi y}{a} + \\ Q\left(\dfrac{m\pi y}{a}\cosh\dfrac{m\pi y}{a} + 3\sinh\dfrac{m\pi y}{a}\right) \end{array} \right] \sin\frac{m\pi x}{a}
$$

(A.24)

$$
\frac{\partial^2}{\partial x^2}\left(\frac{\partial w_1}{\partial y}\right) = -\sum_{m=1,3,5}^{\infty} K\left(\frac{m\pi}{a}\right)^3 \left[\begin{array}{l} J \sinh\dfrac{m\pi y}{a} + \\ Q\left(\dfrac{m\pi y}{a}\cosh\dfrac{m\pi y}{a} + \sinh\dfrac{m\pi y}{a}\right) \end{array} \right] \sin\frac{m\pi x}{a}
$$

(A.25)

$$
\frac{\partial^3 w_1}{\partial y^3} + (2-v)\frac{\partial^3 w_1}{\partial x^2 \partial y} =
$$

$$
\sum_{m=1,3,5}^{\infty} K\left(\frac{m\pi}{a}\right)^3 \left[J\sinh\alpha_m + Q\left(\begin{array}{l} \alpha_m \cosh\alpha_m + 3\sinh\alpha_m - \\ (2-v)(J\sinh\alpha_m + Q(\alpha_m \cosh\alpha_m + \sinh\alpha_m)) \end{array} \right) \right] \sin\frac{m\pi x}{a}
$$

(A.26)

where, $\alpha_m = \dfrac{m\pi a}{2b}$

plug (26) into (2) and solve for Em,
then plug into (18) to have complete equation for w2
Complete equation for w1 in (1)
z = w1+w2, where z is the plate profile as a function of (x,y)

12 Resistor Networks

Resistor networks involve breaking a thermal problem into a combination of resistor-capacitor (R-C) network representations of thermal systems [1–8]. The nodes represent the capacitor and physical represents the discretized mass portions of the model, while the conduction between the nodes are represented by resistors; and the conduction could come from any of the three heat transfer mechanisms—conduction, convection, and radiation. As opposed to other geometry-intensive discretizing methods including finite difference, finite element, and finite volume, R-C networks require minimal computational power, they do not require advanced graphical interfaces, and they are very crash resistant and lend themselves well to parametric studies. The use of a preprocessor that accepts the user's definition of the R-C or flow network parameters and converts user-supplied "Fortran-like" language into real Fortran is the basis of a popular compiler called SindaFluint [sf50]. This Fortran is compiled and sent to the system loader. The loader collects the necessary parts of the precompiled Fortran subroutines that make up the processor library. Because many of the subroutines are invoked through user-supplied calls, each processor load module is essentially custom designed by the user for the problem at hand. When the collect/load task is finished, the resulting load module is executed.

Heat conduction is assumed to be one-dimensional. Fourier's law of heat conduction is:

$$\dot{q}_{cond} = -k \frac{\partial T}{\partial x}$$

(12.1)

Where k is thermal conductivity of the body (W/m/K).

Integrated heating is obtained by integrating the heat flux by forward finite differencing:

$$q^{j+1} = q^j + \dot{q}^j \Delta t$$

(12.2)

Where
q = integrated heating (J/cm^2)
Δt = time incremental step

CONDUCTION

$$\frac{\partial T}{\partial t} = -\frac{1}{\rho c V} \int_s \left[\left(\dot{\vec{q}}_{conv} + \dot{\vec{q}}_{rad} + \dot{\vec{q}}_{cond} \right) \bullet \hat{n} \right] dA$$

(12.3)

DOI: 10.1201/9781003247005-12

The front half is hotter than the back half and radiates heat both outwards and to the back. It is cooled by heat conduction to the back half. Substituting (1) for $\vec{q}_{cond}$ yields:

$$\frac{\partial T}{\partial t} = \frac{1}{\text{thermal mass}}\left[A_{conv}\dot{q}_{ref}HF - A_{rad}\varepsilon\sigma\left(T_3\right)^4 - A_{rad}\varepsilon\sigma\left(\left(T_3\right)^4 - \left(T_4\right)^4\right) - A_{cond}k\frac{\left(T_3 - T_4\right)}{x_{cond}}\right]$$

$$(12.4)$$

Where x_{cond} is the path length of heat conduction and HF is the heating factor of atmospheric entry. Simplifying:

$$\frac{\partial T}{\partial t} = \frac{1}{\text{thermal mass}}\left[A_{conv}\dot{q}_{ref}HF - A_{rad}\varepsilon\sigma\left(2\left(T_3\right)^4 - \left(T_4\right)^4\right) - A_{cond}k\frac{\left(T_3 - T_4\right)}{x_{cond}}\right]$$

$$(12.5)$$

SINDA key points:

- SINDA is a network solver , specifically of R-C networks
- It uses a robust forward-back implicit numerical solving algorithm that is highly stable, O^2 accurate over space and time
- You can insert temperature-dependent properties, such as specific heats, thermal conductivities, and so on. SINDA references these easily with computational robustness.

SINDA R-C network:

- Define thermal energy storage capacity for each node:—thermal mass = rho*vol*c_p = m*c_p (J/K)—SINDA code:
  1,Ti, m_1*c_{p1} $ thermal mass node 1
  2,Ti, m_2*c_{p2} $ thermal mass node 2
  - 10,Ti,0.0 $ node 10 is a boundary node
  Variable geometry cross-sections are not a problem for multi-nodes.
  Radiation conductors, or "rad ks":

- SINDA treats radiation as a radiation conductor.

  - From our previous example $q_{12} = G_{12}(T_2 - T_1)$

    where G_{12} = kA/L between nodes 1 and 2
  For radiation $q = seA(T_2^4 - T_1^4)$, where s is the Stefan-Boltzmann constant, e = emissivity between 1 and 2, and A = area between 1 and 2.
    e*A is also known as the rad k or radiation conductance (W/K⁴)
    In SINDA code:

  - 12,1,2, e*A

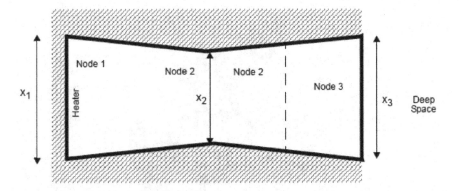

FIGURE 12.1 Bar of variable cross-section.

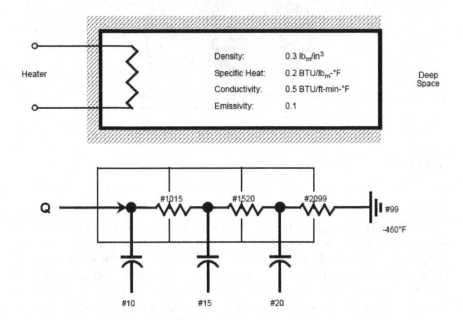

FIGURE 12.2 Bar with internal heating represented by R-C network.

Where e and A are defined between nodes 1 and 2. The negative tells SINDA this is a radiation conductor.

SINDA will automatically solve for q_{12} by multiplying the rad k G_{12} by its calculated temperature difference $T_2^4 - T_1^4$. It does this for each iterative time step.

Simple SINDA example:

- You can model simple cases, such as a simple three-node model (SINDA manual sample problem).

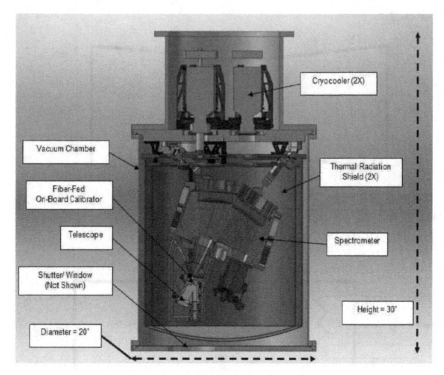

NIS Primary Assembly – CutAway – Components

FIGURE 12.3 NEON spectrometer.

SINDA R-C network tree:

SINDA makes it simple to program the R-C network tree. You need to identify all possible thermal communication branches and it is easily to keep adding nodes and complexity as-needed. SINDA and VBA solve conductivity similarly, but structures problem as a R-C network and allows multi-nodes to be handled with ease. As great as SINDA is, it doesn't prevent anyone from making bad assumptions. Conductances between nodes must be modeled accurately; this is independent of SINDA.

SINDA offers "bells and whistles" that can be called to perform certain tasks (e.g. identify phase change materials, PID controllers, etc.). There were a few assumptions VBA makes that I would like to address. Thermal Desktop offers easy modeling of dynamic environments (conduction and radiation). Convection would have to be inputted in by data base, or by closed form equations and allows ease of creating and analyzing complex geometries.

Thermal Desktop offers easy modeling of dynamic environments (conduction and radiation).). Convection would have to be inputted in by data base, or by closed form equations and allows ease of creating and analyzing complex geometries.

Thermal Desktop, which uses SINDA as the solver, helps with CAD geometry, has a point-and-click GUI, and offers Monte Carlo ray tracing, orbit simulations,

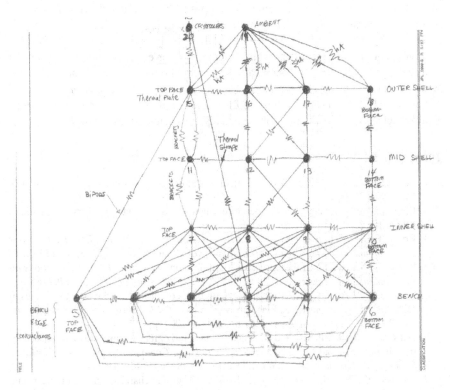

FIGURE 12.4 25-node R-C network tree model with conduction, convection, and radiation.

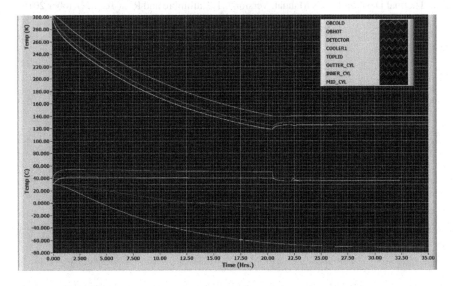

FIGURE 12.5 Transient analysis performed to identify time to cooling of optical bench used for trade-off analysis.

and so on. Ray tracing gives Thermal Desktop the ability to handle view factor calculations much faster than radiosity methods found in finite element solvers. In fact, experience has shown that ray tracing can reduce analysis time over 300%. Solving the complex dynamic view factors during an orbit, for example, could take several hours to solve using finite element radiosity estimations, whereas Monte Carlo ray trace would take only tens of minutes.

Lastly, SINDA has a robust solver that can handle large time steps (Crank-Nicholson), making multiphysics analysis possible in record time.

REFERENCES

1. Vehicle Breakup Analysis (VBA) Software Requirements Specifications, Version 1.0, December 5, 2002.
2. Ling, L., and A. Salama, "Temperature calculations for category I thermal nodes," March 10, 2003.
3. Bosch, E.G.T., "Thermal compact models: An alternative approach," *IEEE Transactions on Components and Packaging Technologies*, vol. 26, no. 1, March 2003, pp. 173–178.
4. Sabry, M.-N., "Compact thermal models for electronic systems," *IEEE Transactions on Components and Packaging Technologies*, vol. 26, no. 1, March 2003, pp. 179–185.
5. Shidore, S., V. Adams, and T.Y.U. Lee, "A study of compact thermal model topologies in CFD for a flip chip plastic ball grid array package," *IEEE Transactions on Components and Packaging Technologies*, vol. 24, no. 2, June 2001, pp. 191–198.
6. Bar-Cohen, A., T. Elperin, and B. Eliasi, "Characterization of chip packages—justification, limitations, and future," *IEEE Transactions on Components, Hybrids, and Manufacturing Technology*, vol. 12, no. 4, December 1989, pp. 724–731.
7. Lasance, C.J.M., H. Vinke, and H. Rosten, "Thermal characterization of electronic devices with boundary condition independent compact models," *IEEE Transactions on Components and Packaging Technologies*, vol. 18, no. 4, December 1995, pp. 723–731.
8. Thermal Desktop User's Manual, Version 6.1, Cullimore and Ring Tech., October 2019.

13 Thermal Analysis Case Studies

INTRODUCTION TO THE CASE STUDIES

The following case studies were meant to demonstrate how to perform analysis for various flight projects and for various flight electronics. The reader should use these as a guide for doing their own analysis. SMC-160C, for example, has qual temperature of 16 °C above the maximum predicted environment (MPE) temperature. Let's say that the MPE is 55 °C on the hot side, then the maximum qual temperature is 71 °C. It is typical to hold the radiator, or the surface where the electronics hardware is temperature controlled, at 71 °C and arrive at a junction temperature through the analysis. Typically, a steady-state analysis is sufficient as a conservative estimate. However, a transient analysis is often required if the environment changes abruptly and quickly, such as during the launch period. Analysis on the cold side of the qualification temperatures is rarely required and usually performed for use in a fatigue or damage analysis at the cold temperature. Some analysis can be strictly performance related, such as understanding the movement of optical mirrors. Also, generally speaking, it is conservative to assume no radiation exchange when performing an analysis at the box level, and most of these analyses assume just that.

INTEROFFICE MEMORANDUM
Project: Mars 2020
February 6, 2019

SUBJECT: THERMAL ANALYSIS OF SCANNER DEMODULATOR ELECTRONICS (SDE) FOR SHERLOC

REF: 1. SMC-160 C. Thermal Requirements
2. ECSS-Q-ST-30–11C, Rev-1. Derating Electronics Specification

CONCLUSION

Q6 and Q7 exceeded the derated guideline of 125 °C for a qual temperature of 70 °C for the baseplate; for a module with thermal grease, the temperature was predicted to be 133.3 °C. Without the thermal grease as planned for flight, the temperature was predicted to be 139.1 °C *Revision Addendum*:

A peer review of this analysis was done by an independent reviewer, and the conclusion was that contact resistance analysis for Q6 and Q7 provided by the vendor was overconservative due to lack of heat spreading allotted in the analysis. When accounting for 30 °C of heat spread from die adhesive, all the way through the hybrid substrate, the contact resistance drops from 961 °C/W, to 254 °C/W. This would put

DOI: 10.1201/9781003247005-13

the junction temperature below the derated value of 125 °C for worst-case qual temperature of 70 °C and no thermal grease between hybrid and lid. Because the thermal spreading cannot be fully estimated without a layout of the hybrid, which there was no access to, the reviewer's analysis should be used to show that the analyst's estimated junction temperatures are conservatively high.

MODEL ASSUMPTIONS

A thermal model of the complete scanner demodulator electronics (SDE) chassis assembly, including both the main PWA and capacitor bank PWAs, was created using SOLIDWORKS Simulation finite element code and composed of 60,744 elements and 106,255 elements. The goal of the analysis was to provide a chassis to hybrid case temperature that would be used to estimate the hottest junction temperatures of the hybrid module. The main PWB was attached to the chassis using 10 perimeter 4–40 screws with a thermal resistance of 2.2 °C/W each per NASA standards specification. The chassis thermal attachment to the 70 °C sink is through mounting screws to the SHERLOC (scanning habitable environments with Raman and luminescence for organics and chemicals) structure.

HYBRID TO LID THERMAL INTERFACE ASSUMPTIONS

The hybrid total power was calculated to be 0.76 W. The PCB had a power distribution of 0.73 W and was smeared across the board of the SDE module. An aluminum-to-aluminum contact conductance was distributed between cover and lid of 480 W/m^2 °C The hybrid has a series of leads that are pressed fit into a receiving connector on the board; this interface was calculated to have a contact resistance of about 1 °C/W. The original design of the hybrid had a thermal grease placed between hybrid surface and lid. There is not a formal gap between hybrid and lid and is metal to metal, though the grease was placed to otherwise make up for air gaps. JPL determined that the grease could not be included in the module and chose to investigate just leaving the metal-to-metal interface, or possibly milling out a pocket over the hybrid and placing a thermal interface pad that is removable and reworkable. All three cases were modeled thermally and compared. The results of the hybrid case temperature for the case where a thermal grease was present showed a maximum case temperature of 72 °C. The condition where there is a metal-to-metal contact only between hybrid and lid had a maximum case temperature of 77.8 °C, and the condition where a Therm-a-gap thermal interface material was placed over the hybrid showed a maximum case temperature of 74.2 °C. The Q6 and Q7 devices have a theta jc value of 80.63 °C/W, which rendered the junction temperatures seen in Table 13.1.

MATERIAL PROPERTIES

Aluminum, 180 W/m-K Polyimide glass, 0.56 W/m-K through thickness (z-direction), 72 W/m-K in-plane (XY plane) solder pads, 50 W/m-K Therm-a-gap 579 thermal impedance@ 10 psi: 0.7 °C-in^2/W Thermal grease: 3.1 W/m-K Kovar, 17 W/mK metal-to-metal Kovar to aluminum: 3.3 °C/W.

TABLE 13.1
Q6 and Q7 Junction Temperature

Thermal Case	Q6, Q7 Junction Temperature, °C
Thermal grease	133.28
Thermal interface pad	135.5
Metal-to-metal only	139.1

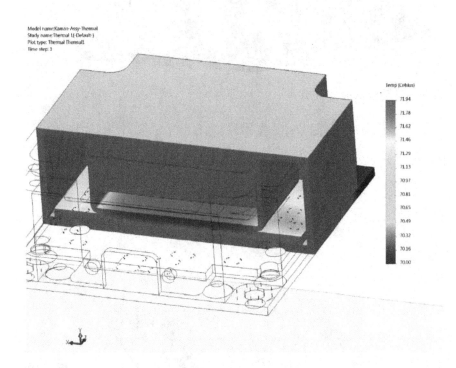

FIGURE 13.1 Maximum hybrid case temperature with thermal grease.

REVISION A—HEAT SPREADING ANALYSIS ON SDE

An independent review of the SDE thermal analysis indicated an overly conservative junction to case thermal resistance of the Q6 and Q7 diodes, a spreadsheet analysis had this thermal resistance as 961 °C/W. In order to obtain better accuracy in the model, a heat spreading analysis was made that simulated how the heat spreads through the molybdenum alloy substrate, hybrid alumina substrate, and Kovar case (Figure 13.4).

Figure 13.5 shows the seven layers that make up the attachment scheme for Q6 and Q7 on the hybrid package. The spreading analysis model included each individual layer.

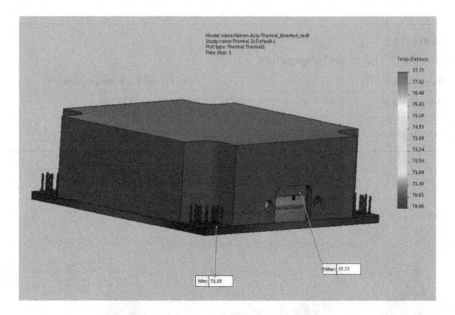

FIGURE 13.2 Maximum case temperature without thermal grease.

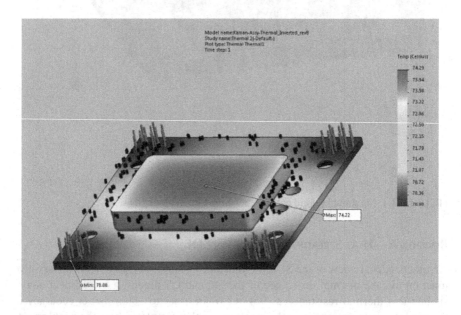

FIGURE 13.3 Hybrid top surface temperature with thermal interface material.

Q6 and Q7 represented the hottest devices on the hybrid as indicated in the SDE thermal analysis. Figure 13.6 shows the scale size of the device in context with the hybrid substrate (blue) and Kovar case (gray). The results of the thermal model are shown in Figure 13.7, and resulted in a Q6,Q7 die temperature of 106.2 °C.

TABLE 13.2
Junction Temperatures for Devices in Hybrid

Ref Des	Material	Length (in)	Width (in)	Thickness (in)	Layer Set	Die Attach %	Power (W)	Tj	Tcase. ˚C
U1	Silicon	0.099	0.095	0.020	Layer Set 1	100%	0.121	77.19	71.8
U2	Silicon	0.099	0.095	0.020	Layer Set 1	100%	0.121	77.19	71.8
U3	Silicon	0.040	0.040	0.023	Layer Set 1	100%	0.015	75.67	71.8
U4	Silicon	0.040	0.040	0.023	Layer Set 1	100%	0.033	80.32	71.8
U5	Silicon	0.040	0.040	0.023	Layer Set 1	100%	0.019	76.71	71.8
U6	Silicon	0.040	0.040	0.023	Layer Set 1	100%	0.015	75.67	71.8
U7	Silicon	0.040	0.040	0.023	Layer Set 1	100%	0.015	75.67	71.8
U8	Silicon	0.110	0.103	0.020	Layer Set 1	100%	0.188	78.78	71.8
CR1	Silicon	0.015	0.015	0.008	Layer Set 1	100%	0.001	73.63	71.8
CR2	Silicon	0.015	0.015	0.008	Layer Set 1	100%	0.024	115.76	71.8
L1	Silicon	0.150	0.120	0.080	Layer Set 1	100%	0.001	71.82	71.8
Q1	Silicon	0.021	0.018	0.010	Layer Set 2	100%	0.002	73.83	71.8
Q2	Silicon	0.020	0.020	0.007	Layer Set 2	100%	0.021	91.97	71.8
Q3	Silicon	0.020	0.020	0.007	Layer Set 2	100%	0.033	103.50	71.8
Q4	Silicon	0.020	0.020	0.007	Layer Set 2	100%	0.021	91.97	71.8
Q5	Silicon	0.020	0.020	0.007	Layer Set 2	100%	0.033	103.50	71.8
06	Silicon	0.020	0.020	0.007	Layer Set 2	100%	0.064	133.28	71.8
Q7	Silicon	0.020	0.020	0.007	Layer Set 2	100%	0.064	133.28	71.8
Q8	Silicon	0.020	0.020	0.007	Layer Set 2	100%	0.005	76.60	71.8

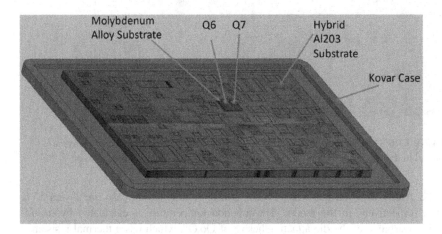

FIGURE 13.4 Hybrid packaging including Q6 and Q7 diodes, molly alloy substrate, hybrid alumina substrate, and Kovar case.

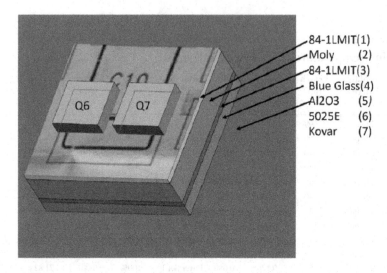

FIGURE 13.5 Adhesive and glass layers associated with Q6 and Q7 diodes.

FIGURE 13.6 Q6,Q7 diode package, hybrid substrate (blue), and Kovar case (gray) are captured in the spreading model, as well as all the layers that make up the stack-up.

The Al203 alumina substrate, 5025E adhesive, and Kovar case are layers 5, 6, and 7, respectively, in the model. The junction temperature is calculated by including the resistance of the die attach adhesive of Q6,Q7, which has a thermal resistance of 4.32 °C/W, and including the thermal resistance between die and junction equal to 4.6 °C/W. The calculation is shown in Figure 13.8 and assumes a margined power of 80 mW for each Q6 and Q7 device, and it shows an estimated junction temperature of 106.91 °C when a worst-case Kovar case temperature of 96.7 °C is assumed.

TABLE 13.3

Board Temperature Results of Select Devices on Main PWB assuming 30 °C Boundary

Reference Designator	Device	Board	Worst Case Power, Steady State (W)	Board Temperature, °C
M1	MOSFET XE "MOSFET"	Main	1.25	70.05
TX1	Magnetic	Main	1.25	66.77
R2	Resistor	Main	0.15	69.43
M4	MOSFET XE "MOSFET"	Main	12.5 @ 50% DC	76.42

The 96.7 °C worst-case Kovar case temperature was obtained from a chassis-level model shown in Figure 13.9. The chassis model contains the SDE package elements including chassis, baseplate, hybrid Kovar case, PCB, and Ultem thermal isolation washers. Figure 13.10 shows the location of the SDE platinum resistance thermometer (PRT). Four different conditions were modeled and the results are located in Table 13.3 at the SDE measured worst-case power of 1.01 W. The 1.01 W total power is a combination of hybrid power of 0.885 W, and PCB power of 0.126 W, per the vendor's test data.

INTEROFFICE MEMORANDUM

Project: Mars 2020

January 31, 2018

SUBJECT: SHERLOC LASER POWER SUPPLY TRANSIENT THERMAL ANALYSIS

REF: 1. SMC-160 C. Thermal Requirements

2. ECSS-Q-ST-30–11C, Rev-1. Derating Electronics Specification

CONCLUSION

The EM002 module with the 12 seconds on/off case saw a maximum MOSFET temperature of 84.6 °C, and the 15 seconds on/off case 85.9 °C at the end of the run. For FM module with thermal adhesive, 12 seconds on/off had a maximum MOSFET temperature of 81.2 °C. For 15 seconds on/off, it had a maximum temperature of 82.8 °C. These model runs assumed an initial and boundary temperature of 30 °C.

MODEL ASSUMPTIONS

A thermal model of the complete laser power supply (LPS) chassis assembly, including both main PWA and capacitor bank PWAs, was created using SOLIDWORKS Simulation finite element code and composed of 60,267 nodes and 29,771 elements.

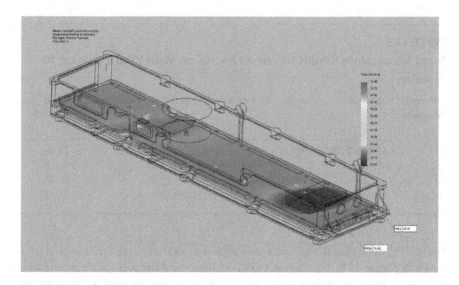

FIGURE 13.7 Thermal gradient of main PWB showing min and max temperatures.

The main PWB was attached to the chassis using 10 perimeter M2 screws with a thermal resistance of 3.1 °C/W each per NASA standards. The PWB is wet-wet-mounted to the chassis with Nusil 2946 thermal bonding adhesive around the entire perimeter. The chassis thermal attachment to the 70 °C sink is through a bracket and a flexure at each end. A total dissipation of 15.15 W was applied to the main board spread across five devices, while the capacitor banks were treated as passive. Gas conduction and radiation at 70 °C were included in the model. Thermal planes were estimated across the entire main board with a focus placed on careful spreading estimation underneath the high power MOSFET devices M3 and M4.

A transient analysis was performed of M3 and M4 to assure that a duty cycle of 50% of the peak power estimate was adequate. The TX1 inductor device was assumed to conduct heat away by means of an M2 screw to the chassis, while the rest of the devices conducted heat through solder pads into the main PWB.

MATERIAL PROPERTIES USED

Aluminum, 180 W/m-K
   Polyimide-glass, 0.56 W/m-K through thickness (z-direction), 72 W/m-K in-plane (XY plane)
   Solder pads, 50 W/m-K
   PWB to chassis adhesive resistance (Nusil CV 2946), 0.103 °C/W (assume 30% void)

THERMAL RESULTS

Figure 13.5 shows the results of the thermal analysis, with a hottest temperature of 76.42 °C measured underneath M4.

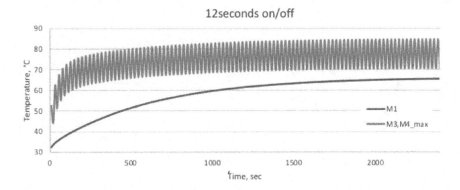

FIGURE 13.8 Temperature vs. time results for maximum MOSFET temperature of M3 and M4, and M1 temperature for 12 seconds on, 12 seconds off.

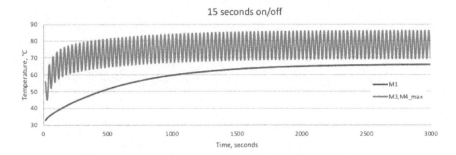

FIGURE 13.9 Temperature vs. time results for maximum MOSFET temperature of M3 and M4, and M1 temperature for 15 seconds on, 15 seconds off.

No Thermal Adhesive (EM 002)

The plot shown in Figure 13.5 shows the temperature results for the following conditions assuming no thermal adhesive.

DETAILED MODE (100 SPECTRA)

Pulses = 900, Frequency = 80 Hz, Width = 40 usec
Current = 25A, Chassis temperature = 30 °C, 12 seconds on/off
TX1 = 1.25 W. Run to steady state

Temperature versus time results are shown in Figure 13.8 for the following conditions assuming no thermal adhesive:

Detailed Mode (100 spectra)
Pulses = 900, Frequency = 80 Hz, Width = 40 usec
Current = 25A, Chassis temperature = 30 °C, 15 seconds on/off

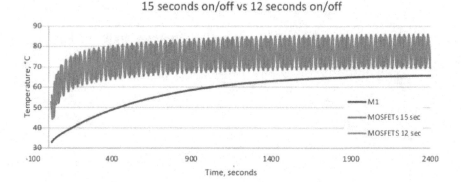

FIGURE 13.10 Temperature vs. time results for maximum MOSFET temperature of M3 and M4, and M1 temperature for overlapped 12 seconds and 15 seconds plot.

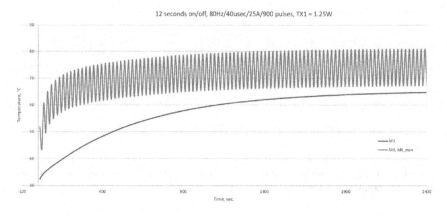

FIGURE 13.11 Temperature vs. time results for maximum MOSFET temperature of M3 and M4, and M1 temperature for 12 seconds on, 12 seconds off with thermal adhesive.

TX1 = 1.25 W
Run to steady state

Overlap of 15 seconds and 12 seconds plots are shown in Figure 13.5 for no thermal adhesive.

With thermal adhesive (FM)

Detailed Mode (100 spectra)
Pulses = 900, Frequency = 80 Hz, Width = 40 usec
Current = 25A
Chassis temperature = 30 °C, 12 seconds on/off
TX1 = 1.25 W
Run to steady state

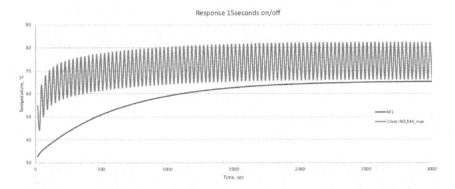

FIGURE 13.12 Temperature vs. time results for maximum MOSFET temperature of M3 and M4, and M1 temperature for 15 seconds on, 15 seconds off with thermal adhesive.

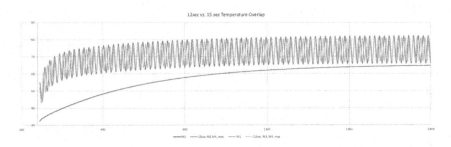

FIGURE 13.13 Temperature vs. time results for maximum MOSFET temperature of M3 and M4, and M1 temperature for 12 seconds and 15 seconds plots overlapped, with thermal adhesive.

Detailed Mode (100 spectra)
Pulses = 900, Frequency = 80 Hz, Width = 40 usec
Current = 25A
Chassis temperature = 30 °C, 15 seconds on/off
TX1 = 1.25 W
Run to steady state

The thermal plot in Figure 13.13 shows overlapped of 15 seconds and 12 seconds plots, with thermal adhesive.

TRANSIENT TEMPERATURE OF MOSFETs vs. CHASSIS TEMPERATURE (TC 14 LPS) FOR EM 002

Figure 13.14 shows the temperature versus time of the TC 14 LPS chassis and the maximum MOSFET temperature; this is for the EM (no thermal goop) module. This was assuming an initial temperature of 30 °C. We can expect a maximum temperature of about 82.4 °C as shown by the MOSFET, which will be 56.4 °C for location TC 14 LPS chassis.

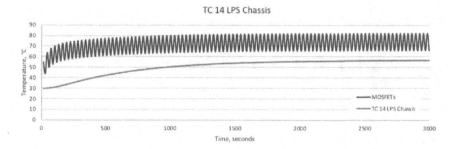

FIGURE 13.14 Temperature vs. time results for maximum MOSFET temperature of M3 and M4, and chassis temperature (TC 14) for 15 seconds on, 15 seconds off with thermal adhesive.

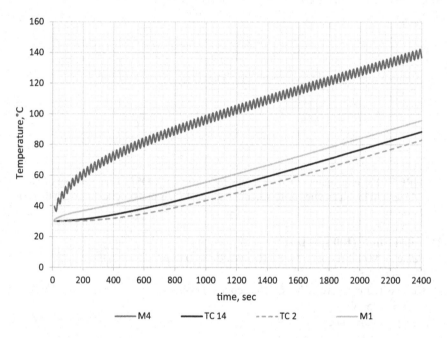

FIGURE 13.15 Adiabatic temperature vs. time results for maximum MOSFET temperature M4, and chassis temperature (TC 14) for 12 seconds on, 12 seconds, no thermal adhesive. M4 will reach 100 °C at approximately 972 seconds (16.2 minutes).

Adiabatic Transient Temperature of MOSFETs vs. Chassis Temperature (TC 14, TC 2 LPS) for EM 002

The transient analysis was run with the 30 °C boundary temperature removed at the LPS mounting locations, essentially cutting off the ability for the LPS to conduct, convect, or radiate heat to its surroundings. Figure 13.15 shows the results of the plot with the temperature of thermocouples TC 2 and TC 14.

U1—THERMAL STRAP THERMAL ANALYSIS
AT 55 °C BOUNDARY TEMPERATURE

A 2 mm copper strap was attached to U1 in order to reduce the maximum temperature as shown in Figure 13.16. The benefit illustrated in Figure 13.17 shows a 20-degree drop in maximum case temperature of U1 with the addition of the copper strap. The transient plot was run with a 45 °C interface temperature.

The following analysis runs were done assuming 55 °C flight acceptance temperature for testing. The bolted interfaces of the LPS have thermal isolators and are assumed set at 55 °C. Figure 13.18 shows the condition where only quiescent power is turned on and set at 6.6 W total, where 4.4 W are applied at U1 and 2.2 W is spread across the board. The maximum temperature is 84.61 °C at U1. Figure 13.19 shows the temperatures while pulsing. The MOSFET maximum temperature occurs at M4 and is 97.3 °C.

INTEROFFICE MEMORANDUM

Project: NISAR

March 23, 2017

SUBJECT: UNIVERSAL SYSTEM TRANSPONDER
(UST) THERMAL ANALYSIS

REF: 1. SMC-160 C. Thermal Requirements

        2. ECSS-Q-ST-30–11C, Rev-1. Derating Electronics Specification

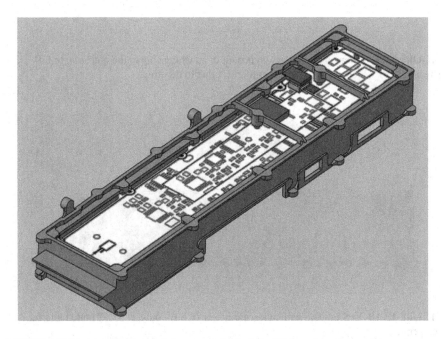

FIGURE 13.16 A copper thermal strap was attached the U1 to reduce the package temperature.

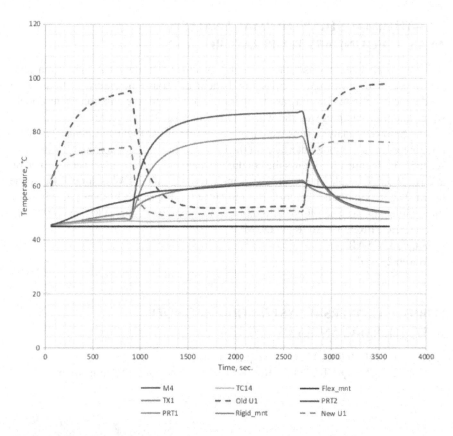

FIGURE 13.17 Temperature reduction benefit of attaching copper thermal strap to U1 shwoing a 20 °C reduction in temperature of U1 due to the strap.

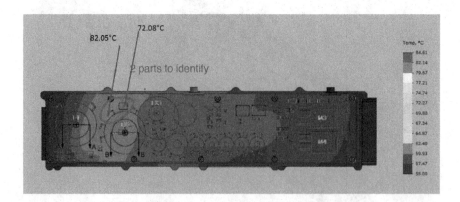

FIGURE 13.18 LPS module with quiescent power (6.6 W) only, at 55 °C bolted interface temperature.

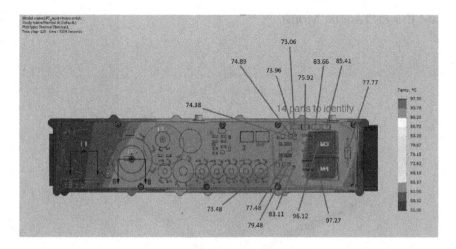

FIGURE 13.19 shows 14 parts temperatures for pulsing 15.15 W case at worst-case hot at 55 °C bolted interface temperature, where the two MOSFETs M3 and M4 dissipate 6.75 W each.

Background:

A steady-state thermal analysis has been completed on the KaM chassis circuit card assembly. The total power of the KaM was simulated at 45.6 watts. The thermal simulation assumed a digital power module (DPM) interface temperature of 70 °C, and the resultant chassis level thermal gradient was used as a boundary condition for the individual slice assemblies.

Conclusion:

1. On the SPM, the simulation showed the following exceeded ECSS derated junction temperature of 85 °C: U300 (FPGA) reached 90.7 °C, U6,U8 reached 90.8 °C, and U16 reached 103.8 °C.
2. On the GPSM, no device exceeded the JPL derated junction temperature.
3. On KTM, the simulation showed the following exceeded JPL derated junction temperature of 85 °C: U11 reached 92.7 °C, U7 reached 98.4 °C, U43 reached 91.7 °C, and U38. The following exceeded ECSS derated junction temperature of 110 °C: U12 reached 107.6 and U19 reached 160.2 °C. U9 had an ECSS derated junction temperature of 95 °C; the simulation showed it reached 125.7 °C.

Discussion:

A steady-state FEM thermal analysis was performed using a finite element model (FEM) created in SOLIDWORKS Simulation. SOLIDWORKS Simulation offers the ability to run thermal analyses with the existing mechanical CAD geometry allowing for accuracy and relatively quick simulation turnaround times. Five simulations were performed for the following subassemblies: GPSM, SPM, PSM, KTM, and chassis. The KaM chassis model was run in order to establish boundary conditions for the individual PWA assembly simulations.

KAM CHASSIS THERMAL MODEL

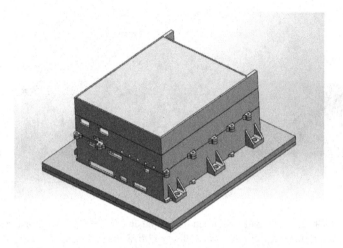

FIGURE 13.20 KaM chassis assembly.

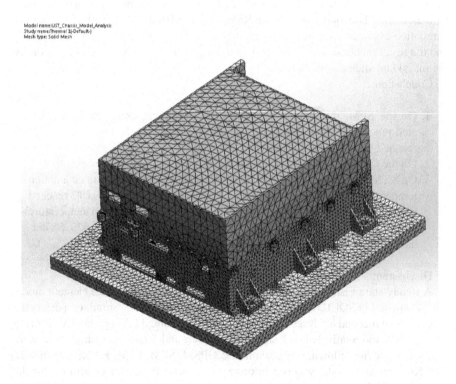

FIGURE 13.21 91k node thermal model.

TABLE 13.4
Materials Properties Used in the Analysis

Material	k (W/m°C)
Chassis (Al 6061)	169.6
Polyimide	0.4
R04350B	0.69
Copper	390
Alumina-Solithane	0.64
Kovar	17

MODEL ASSUMPTIONS AND BOUNDARY CONDITIONS

The KaM chassis thermal model consisted of all three KaM subchassis (DPM, PSM, and KTM), as well as the SPM and GPSM PWBs. Power was distributed normally on the PWB for GPSM and SPM, but for the KTM and PSM, power was distributed along the PWA screw-down locations. M6 screws (0.7 °C/W) were assumed at the base of the DPM dry-mounted to a k-core spreader plate. There is an assumed wet-mount interface between k-core spreader and radiator. The material has a thermal conductivity of 5 W/m °C with a thickness of .005 inch. A bond thermal conductance of 450 W/m^2-K was assumed between DPM/PSM and PSM/KTM chassis interfaces. Thermal conductance was used instead of fastener resistances in part because the screws are not underneath the contact area between the chassis, and are more off to the side as tabs, or "ears." 450 W/m^2-K is typically used for conservative dry-mount chassis mounting. All chassis were made of Al 6061-T6. Radiation was neglected in this analysis.

MATERIALS PROPERTIES

The materials properties used in the analysis are shown in Table 13.4.

PWA MODELS

For the PWB laminate thermal conductivity values, I used orthotropic values derived by the board stack-up for each PWB. The effective thermal conductivities both in-plane of the PWB (in the XY plane of the board) and in the out-plane transverse direction (z-direction) are as follows:

TABLE 13.5

PWB Orthotropic Thermal Conductivity Values

PWB	k_{in}, W/m °C	k_{out}, W/m °C
GPSM (10390662 rev A)	69.39	0.59
SPM (10390672 rev A)	61.47	0.95
KTM (10390912 rev A)	29.68	0.75
FRM (10390902 rev A)	25.20	1.00

$$k_{in-plane} = \frac{\sum_{1}^{N} \eta_i k_i t_i}{\sum_{1}^{N} t_i}$$

$$k_{out-plane} = \frac{\sum_{1}^{N} t_i}{\sum_{1}^{N} \frac{t_i}{\eta_i k_i}}$$

Where k_i = thermal conductivity of the ith layer,
t_i = thickness of ith layer,
η_i = % copper of the ith signal, power, or ground layer (correction factor)

The results are tabulated in Table 13.5.

The out-of-plane thermal conductivity of the SPM was augmented to reflect thermal vias under hot components, or 0.95 W/m°C, changed from 0.56 W/m°C without thermal vias.

DPM SLICE

SPM
The SPM PWA model consisted of 21,234 total nodes and 10,407 elements. The board was meshed with tetrahedral elements and component power was applied underneath each components. Thermal resistance between component and PWB was applied corresponding to the package type. The boundary temperatures at the chassis screw locations were obtained from the KaM chassis thermal model. M3 screws were assumed with a thermal resistance of 3 °C/W per screw.

Virtex 5—The FPGA component was treated and modeled as a flip chip on alumina substrate and CCGA connections to the board. A thermal lid was also modeled with an attached .040-inch copper strap.

The GPSM PWA model consisted of 20,122 total nodes and 9,837 elements. The board was meshed with tetrahedral elements and component power was applied

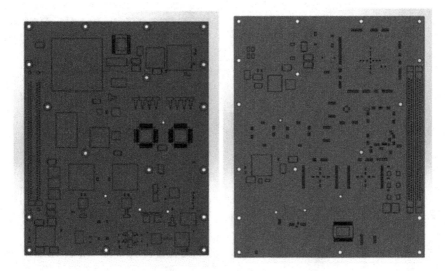

FIGURE 13.22 SPM PWA component side (left) and back side (right) layout.

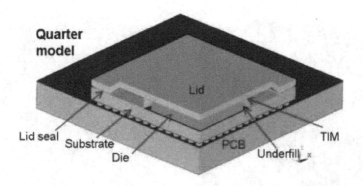

FIGURE 13.23 FPGA stack-up model.

underneath each component. Thermal resistance between component and PWB was applied corresponding to the package type. The boundary temperatures at the chassis screw locations were obtained from the KaM chassis thermal model. M3 screws were assumed with a thermal resistance of 3 °C/W per screw.

PSM SLICE

The PSM chassis assembly and PWA model consisted of 25,244 total nodes and 12,681 elements. The board was meshed with tetrahedral elements and component power was applied underneath each component. The PSM PWA was modeled directly with the chassis due to the thermal importance of the chassis geometry. Thermal

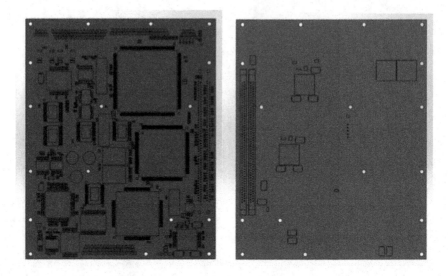

FIGURE 13.24 GPSM PWA component side (left) and back side (right) layout.

FIGURE 13.25 PSM chassis assembly.

resistance between components, PWAs, and chassis was applied corresponding to the package type. The boundary temperatures at the chassis screw locations were obtained from the KaM chassis thermal model. M3 screws were assumed with a thermal resistance of 3 °C/W per screw.

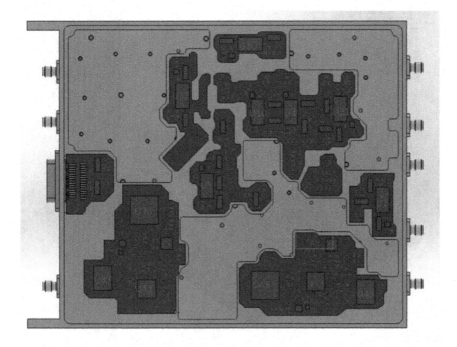

FIGURE 13.26 KTM chassis and PWA assembly. The chassis and PWB were modeled together due to complex geometry.

KTM

The KTM PWA model consisted of 16,688 total nodes and 8,016 elements. The board was meshed with tetrahedral elements and component power was applied underneath each component. The KTM PWA was modeled directly with the KTM chassis due to the complexity of the chassis. Thermal resistance between component and PWB was applied corresponding to the package type. The boundary temperatures at the chassis screw locations were obtained from the KaM chassis thermal model. M3 screws were assumed with a thermal resistance of 3 °C/W per screw.

RESULTS

KaM Chassis

The KaM chassis thermal model showed the thermal gradient to be 70–87 °C, nonuniform at the interfaces between the three chassis slices (Figure 13.27). These results were used as boundary conditions for all the PWA analyses.

SPM PWA

The SPM contains the highest power component (FPGA) but is kept relatively cool by the copper thermal strap that draws most of the heat through the top of the component and off the side chassis wall. The hottest portion of the board occurs at

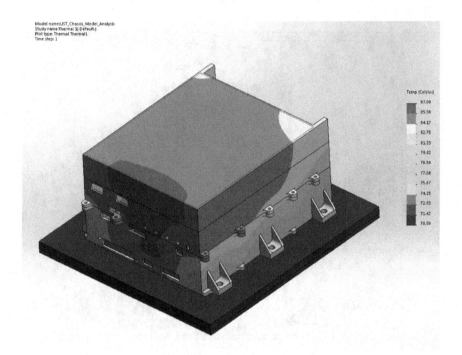

FIGURE 13.27 KaM chassis thermal model showing expected gradient across all the individual chassis. The KaM chassis model contained a kcore spreader plate (shown all blue) fastened with M6 screws at six locations.

the center in part due to the DPM chassis temperature being highest at the screw locations associated with U3 and U4.

GPSM PWA

The GPSM has the hottest part of the board at 92.8 °C in part due to components U102 and U32 located back to back on the PWB toward the center where the DPM chassis is warmest. These components are well below their derated junction temperatures.

PSM Chassis

The PSM was hottest at the location of two adjacent DC/DC converters and stayed relatively cool at the location of the PSM PWA and the FRM PWA.

KTM PWA

The KTM benefited from the large amount of PWB tie-down screws that help bring the temperature down.

INTEROFFICE MEMORANDUM

Project: Europa Clipper

October 4, 2019

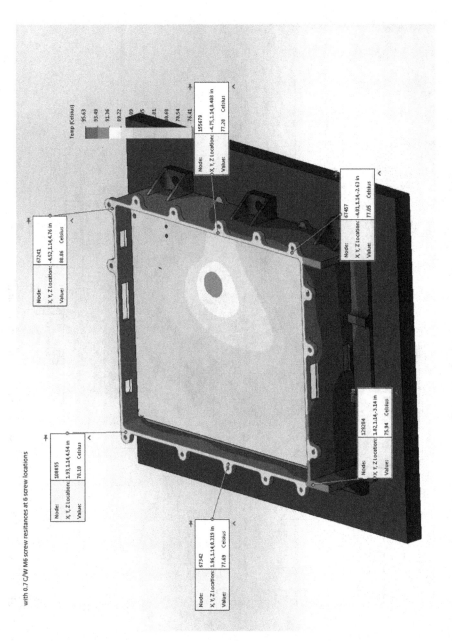

FIGURE 13.28 DPM chassis with GPSM (shown) and SPM (hidden) PWAs.

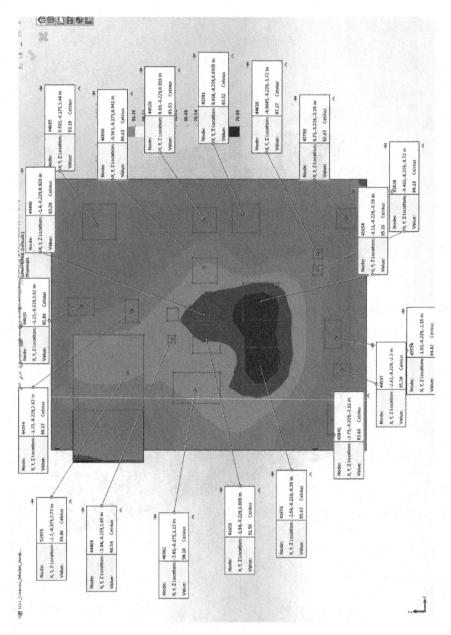

FIGURE 13.29 SPM PWA model. Two center hot spots are coincident with U3 and U4 locations.

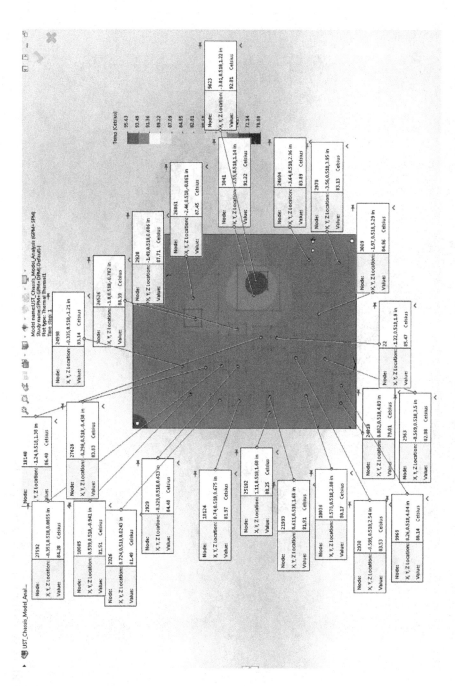

FIGURE 13.30 GPSM PWA model. The hot spot occurs at U102 and U32 are about 1 watt each, which are placed back to back (i.e., opposite sides of the PWB).

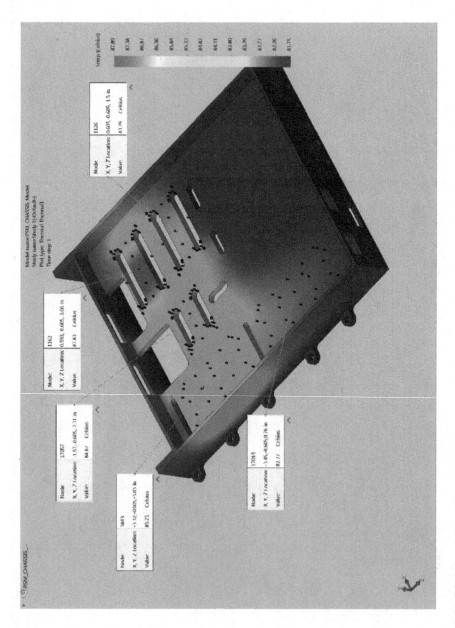

FIGURE 13.31 PSM chassis thermal model results.

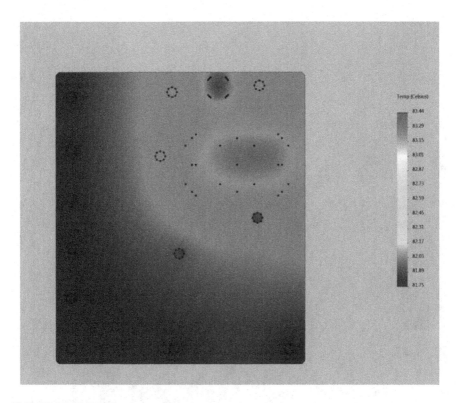

FIGURE 13.32 FRM pad temperatures.

SUBJECT: PRELIMINARY POWER SWITCH SLICE (PSS) ASSEMBLY THERMAL ANALYSIS

REF: 1. SMC-160 C. Thermal Requirements
2. ECSS-Q-ST-30–11C, Rev-1. Derating Electronics Specification

CONCLUSION

For the digital PWA, part U250 had a pad temperature of 94.71 °C, which resulted in a junction temperature of 108.2 °C. All components on the digital PWA met the ECSS derated guideline. For the switch PWA, channel 10A MOSFET, a.k.a FET, was the hottest device and had a pad temperature of 86.7 °C with a resultant junction temperature of 93.4 °C, which was below the ECSS derated guideline of 110 °C. All other components met the ECSS derated guideline on the switch PWA.

MODEL ASSUMPTIONS

The PWBs were assumed to have thermal orthotropic properties, and their values were calculated based on the stack-up provided in the PWB drawing (see Tables 13.11

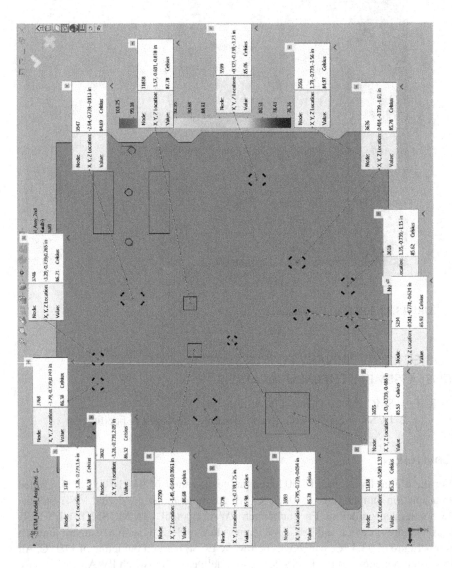

FIGURE 13.33 KTM thermal model results.

TABLE 13.6

GPSM Thermal Analysis Junction Temperature Results (Tjunction) vs. ECSS Derated Junction Temperatures (Derated Tjmax)

Ref.	Package	Power (W)	Pad Temp. °C	Board-Case °C/W	T case °C	Theta JC °C/W	T junction °C	Case to PCB Path	Derated Tjmax, °C
U100	QFP	0.531	91.22	0.86	91.68	2.5	93.00	Al2O3 Solithane	110
U101	QFP	0.33	84.96	0.42	85.10	3.8	86.35	Al2O3 Solithane	110
U2	DFP	0.0033	85.47	125.44	85.88	1.27	85.89	Leads	110
U6, U7	DFP	0.066	84.48	62.03	88.57	22	90.03	Leads	110
U102	QFP	1.08	87.45	0.15	87.61	0.2	87.83	Al2O3 Solithane	95
U104, U22	SOP	0.1	86.4	39.43	90.34	5	90.84	Leads	110
U16	QFP	0.005	80.17	6.36	80.20	10	80.25	Leads	110
U14	DFP	0.1	81.91	155.08	97.42	22	99.62	Leads	110
U13	DFP	0.095	80.25	158.37	95.29	2.44	95.53	Leads	110
U501	DFP	0.165	81.51	1.25	81.72	7	82.87	Al2O3 Solithane	110
U17	DFP	0.066	80.14	81.06	85.49	50	88.79	Leads	110
U67, U29	DFP	0.033	79.01	134.81	83.46	20	84.12	Leads	110
U20	DFP	0.0033	83.89	243.92	84.69	15	84.74	Leads	110
U25	DFP	0.033	84.28	134.81	88.73	20	89.39	Leads	110
U3	DFP	0.066	83.14	122.61	91.23	10	91.89	Leads	110
U1, U4, U30	DFP	0.083	83.53	62.03	88.68	22	90.50	Leads	110
U5	DFP	0.24	83.13	1.53	83.50	9	85.66	Al2O3 Solithane	110
U24	DFP	0.278	92.81	1.53	93.23	9	95.74	Al2O3 Solithane	110
U32	DFP	0.9	88.39	1.53	89.76	9	97.86	Al2O3 Solithane	110
	4.5426								

and 13.12). The digital board had an assumed thermal conductivity of 68.8 W/m-K in-plane, 0.48 W/m-K transverse. The switch board thermal conductivity was 45.4 W/m-K in-plane, and 0.95 W/m-K transverse. A thermal model of the complete power switch slice (PSS) slice assembly, including both digital PWA and switch PWAs, was created using SOLIDWORKS Simulation finite element code. An array of thermal vias were subsequently added below the chassis frame. The dimensions of the vias assumed were .026 in. diameter, with 2 mil of plating and 0.118 in center-to-center spacing.

TABLE 13.7

SPM Thermal Analysis Junction Temperature Results (Tjunction) vs. ECSS Derated Junction Temperatures (Derated Tjmax). (Values that exceeded the ECSS derated Tjmax are highlighted in yellow.)

Ref.	Package	Power (W)	Pad Temp. °C	Board-Case °C/W	T case °C	Theta JC °C/W	T junction °C	Case to PCB Path	Derated Tjmax, °C
U300	CGA	8.08	89.86			0.1	90.67	Leads + Heatstrap to chassis wall	85
U26	DFP	0.08	91.56	82.60	98.17	11.35	99.08	Leads	110
U1, U17	DFP	0.25	92.29	9.69	94.71	50	107.21	Al2O3 Solithane	110
U13, U180	DFP	0.066	85.98	62.03	90.07	22	91.53	Leads	110
U18	QFP	0.33	90.1			not populate		Leads	95
U6, U8	QFP	0.416	89.61			2.96	90.84	Soldered paddle	85
U14	DFP	0.066	88.27	122.61	96.36	10	97.02	Leads	110
U15	Custom	0.375	82.63			66	107.38	Soldered paddle	110
U9, U16	Custom	0.264	88.19			59	103.77	Soldered paddle	85
U19	DFP	0.1	85.3	155.08	100.81	22	103.01	Leads	110
U3, U4	CGA	1.687	95.63			15	120.94	CGA	110
U7	DFP	0.36	83.66	1.53	84.21	9	87.45	Al2O3 Solithane	110
U5	DFP	0.224	83.52	1.53	83.86	9	85.88	Al2O3 Solithane	110
U2	QFP	0.7	83.19	2.00	84.59	1.5	85.64	Al2O3 Solithane	110
		15.265							

TABLE 13.8

PSM Thermal Analysis Case Temperature Results

Ref.	Power (W)	Chassis Temp (°C)	Case Temp (°C)
	8.33	85.21	85.38
	0.06	84.87	84.87
	5.88	87.83	88.20
	2.5	82.77	82.82
	16.77		

TABLE 13.9

KTM Thermal Analysis Junction Temperature Results (Tjunction) vs. ECSS Derated Junction Temperatures (Derated Tjmax). (Values that exceeded the ECSS derated Tjmax are highlighted in yellow.)

Ref.	Package	Power (W)	Pad Temp. °C	Board-Case °C/W	T case °C	Theta JC °C/W	T junction °C	Case to PCB Path	Tjmax Derated °C
U3	Leaded device	0.21	85.3			99	106.09	by soldering pads	110
U1	SMD QFP	0.168	85.2			8.3	86.59	by soldering pads	110
U11	CQFP	0.2025	85.98			33.4	92.74		85
U7	10 lead CERPAC	0.24	86.68	8.97	88.83	40	98.43	Solithane	85
U9	Plastic QFN	1.65	87.78			23	125.73	by soldering pads	95
U12	Leaded device	0.21	86.78			99	107.57	by soldering pads	110
U4	SMD QFP	0.1725	88.05			8.3	89.48	by soldering pads	110
U5	SMD QFP	1.488	101.25			8.3	113.60	by soldering pads	110
U10	SMD QFP	0.03	86.42			8.3	86.67	by soldering pads	110
U19	SMD QFP	0.225	84.69			212	132.39	by soldering pads	110
U19	SMD QFP	0.22	84.69			343	160.15	by soldering pads	110
U19	SMD QFP	0.315	84.69			48	99.81	by soldering pads	110
U15-U18	SMD QFP	0.3	86.38			164	135.58	by soldering pads	110
U30	SMD QFP	0.205	87.81			8.3	89.51	by soldering pads	110
U31	SMD QFP	0.15	87.59			8.3	88.84	by soldering pads	110
U32	SMD QFP	0.44	89.23			8.3	92.88	by soldering pads	110
U43	CQFP	0.2025	84.93			33.4	91.69		85
U38	10 lead CERPAC	0.18	85.19	8.97	86.80	40	94.00	Solithane	85
U36	Leaded device	0.15	85.34			87	98.39	by soldering pads	110
U39	Leaded device	0.21	83.23			99	104.02	by soldering pads	110
U33	SMD QFP	0.09	85.75			8.3	86.50	by soldering pads	110
U34	SMD QFP	0.1725	84.97			8.3	86.40	by soldering pads	110

TABLE 13.9
KTM Thermal Analysis Junction Temperature Results (Tjunction) vs. ECSS Derated Junction Temperatures (Derated Tjmax).
(Values that exceeded the ECSS derated Tjmax are highlighted in yellow.) (continued)

Ref.	Package	Power (W)	Pad Temp. °C	Board-Case °C/W	T case °C	Theta JC °C/W	T junction °C	Case to PCB Path	Tjmax Derated °C
U35	SMD QFP	0.288	85.06			8.3	87.45	by soldering pads	110
U53	CQFP	0.2025	85.22			33.4	91.98		110
U50	10 lead CERPAC	0.18	85.24	8.97	86.85	40	94.05	Solithane	110
U48	Leaded device	0.15	85.44			87	98.49	by soldering pads	110
U51	Leaded device	0.21	85.01			99	105.80	by soldering pads	110
U45	SMD QFP	0.09	85.33			8.3	86.08	by soldering pads	110
U46	SMD QFP	0.1725	85.97			8.3	87.40	by soldering pads	110
U47	SMD QFP	0.288	85.62			8.3	88.01	by soldering pads	110
	Total	8.812							

TABLE 13.10

FRM Thermal Analysis Junction Temperature Results (Tjunction) vs. ECSS Derated Junction Temperatures (Derated Tjmax). (Values that exceeded the ECSS derated Tjmax are highlighted in yellow.)

Ref.	Package	Power (W)	Pad Temp. °C	Board-Case °C/W	T case °C	Theta JC °C/W	T junction °C	Case to PCB Path	Derated Tjmax, °C
U1	QFP	0.06	83.17	328.08	102.86	2.5	103.01	Lead	110
U2		0.21	83.44	3.91	84.26	60	96.86	Solithane/ G10/2216	110
	Total	0.27							

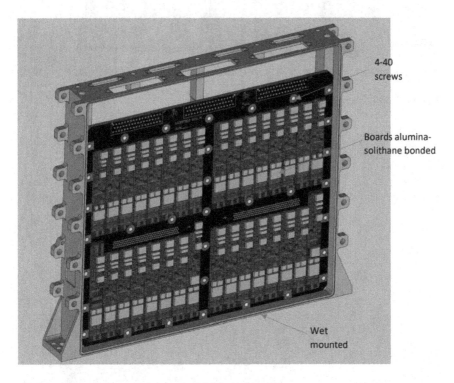

FIGURE 13.34 Thermal model for the PSS switch board.

Figure 13.34 shows how the PWAs attached to the chassis using 4–40 perimeter screws with a thermal resistance of 3.5 °C/W each. The chassis is wet-mounted to the chassis with alumina-solithane bonding adhesive at the cold mounting surface. The chassis thermal attachment temperature is 70 °C. There was a total dissipation

FIGURE 13.35 Thermal model of the digital boards.

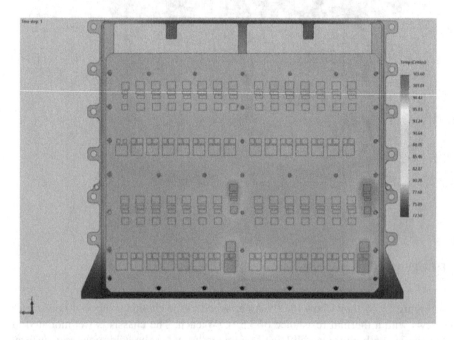

FIGURE 13.36 Switch PWA temperatures with the addition of thermal vias beneath chassis rails.

of 55 W for the switch board and 16.42 W for the two digital boards, or 28.1 W each. Radiation was neglected in the model.

Figure 13.35 shows the topside of the digital boards, left and right, and Figure 13.36 shows the backside, with the switch board hidden. The digital board connected the switch board in a daughter-mother card arrangement, and component power wise and placement wise are identical to each other. The thermal model was run with all three PWBs, one switch and two digital, attached to the chassis.

MATERIAL PROPERTIES USED

Aluminum, 167 W/m-K Solder pads, 50 W/m-K Chassis wet-mount adhesive (2216), 0.22° W/m-K (assume 30% void) Alumina-solithane, 0.64 W/m-K.

THERMAL RESULTS

Table 13.13 shows the results of the thermal analysis of the switch board. The hottest location on the board, at the upper part, was 99.5 °C. The upper part of the chassis was 103.6 °C due to power dissipation at the connectors, but now devices are located on the chassis. No components exceeded the ECSS derated 110 °C junction temperature requirement. The addition of thermal vias beneath the chassis rails and location of the rails close to the hot channel 10A FETs effectively lowers the maximum temperature.

Table 13.1 shows the results of the digital board, showing a maximum case temperature of 94.7 °C for U250, which resulted in a junction temperature of 108.2 °C.

TABLE 13.11

Orthotropic Thermal Conductivity Calculations for Switch PWB

PSS Switch Board

CU Layer		Cu oz./mil	Type	thick (in.)	thick (m)	k, (W/m-K)	Coverage μ_i	(k_{in}, top)	(k_{in}, bot)	(k_{out}, bot)
CU	1	1		0.0014	0.000036	390	0.3	0.00416052	0.00003556	3.03932E-07
Core		4	Core	0.004	0.000102	0.46	1	0.000046736	0.0001016	0.00022087
CU	2	0.5	GND	0.0007	0000018	390	0.6	0.00416052	0.00001778	7.59B29E-08
preg		4	preg	0.004	0.000102	0.46	1	0.000046736	0.0001016	0.00022087
CU	3	0.5	Core	0.0007	0.000018	390	0.95	0.0065 8749	0.00001778	4.79892E-08
Core		4	Core	0.004	0.000102	0.46	1	0.000046736	0.0001016	000022087
CU	4	0.5	CU	0.0007	0.000018	390	0.25	0.00173355	0.00001778	1.82359E-07
preg		4	preg	0.004	0.000102	0.46	1	0.000046736	0.0001016	0.00022087
CU	5	0.5	GND	0.0007	0000018	390	0.9	0.00624078	0.00001778	5.06553E-08
Core		4	Core	0.004	0.000102	0.46	1	0.000046736	0.0001016	0.00022087
CU	6	0.5	GND	0.0007	0.000018	390	0.95	0.0065 8749	0.00001778	4.79B92E-08
preg		4	preg	D.004	0.000102	0.46	1	0.000046736	0.0001016	0.00022087
CU	7	2	GND	0.0028	0000071	390	0.6	0.01664208	0.00007112	3.03932E-07
Core		4	Core	0.004	0.000102	0.46	1	0.000046736	0.0001016	0.00022087
CU	8	0.5	GND	0.0007	0.000018	390	0.95	0.00658749	0.00001778	4.79892E-08
preg		4	preg	0.004	0.000102	0.46	1	0.000046736	0.0001016	000022087
CU	9	2	GND	0.0028	0.000071	390	0.95	0.02634996	0.00007112	1.9195TE-07
Core		4	Core	0.004	0.000102	0.46	1	0.000046736	0.0001016	0.00022087
CU	10	2	GND	0.0028	0000071	390	0.95	0.02634996	0.00007112	1.9195TE-07
preg		4	preg	0.004	0.000102	0.46	1	0.000046736	0.0001016	0.00022087
CU	11	0.5	GND	0.0007	0.000018	390	0.95	0.00658749	0.00001778	4.79892E-08
Core		4	Core	0.004	0.000102	0.46	1	0.000046736	0.0001016	000022087
CU	12	2	GND	0.0028	0.000071	390	0.73	0.020247864	0.00007112	2.49BD7E-07

preg		4	preg	0.004	0.000102	0.46	1	0.000046736	0.0001016	0.00022087
CU	13	2		0.0028	0000071	390	0.3	0.00832104	0.00007112	6.07863E-07
Core		4	Core	0.004	0.000102	0.46	1	0.000046736	0.0001016	0.00022087
CU	14	0.5	Sig	0.0007	0.000018	390	0.3	0.00208026	0.00001778	1.51966E-07
Core		4	Core	0.004	0.000102	0.46	1	0.000046736	0.0001016	000022087
CU	15	2	GND	0.0028	0.000071	390	0.6	0.01664208	0.00007112	3.03932E-07
preg		4	preg	0.004	0.000102	0.46	1	0.000046736	0.0001016	0.00022087
CU	16	0.5		0.0007	0000018	390	0.95	0.00658749	0.00001778	4.79B92E-08
Core		4	Core	0.004	0.000102	0.46	1	0.000046736	0.0001016	0.00022087
CU	17	0.5	Sig	0.0007	0.000018	390	0.25	0.00173355	0.00001778	1.82359E-07
preg		4	preg	0.004	0.000102	0.46	1	0.000046736	0.0001016	000022087
CU	18	0.5	GND	0.0007	0.000018	390	0.9	0.00624078	0.00001778	5.06553E-08
Core		4	Core	0.004	0.000102	0.46	1	0.000046736	0.0001016	0.00022087
CU	19	0.5	Sig	0.0007	0.000018	390	0.3	0.00208026	0.00001778	1.51966E-07
Core		4	Core	0.004	0.000102	0.46	0.3	1.40208E-05	0.0001016	0000736232
CU	20	0.5	Core	0.0007	0.000018	390	0.3	0.00208026	0.00001778	1.51966E-07
Core		4	Core	0.004	0.000102	0.46	0.3	1.40208E-05	0.0001016	0.000736232
CU	21	1	Sig	0.0014	0.000036	390	0.3	0.00416052	0.00003556	3.03932E-07
							Summation	0.183030724	0.00262382	0.005451811

copper	0.0287
dielectric	0.08
top+bot Solder	0
top+bot SM	0
TOTAL	0.1087

TABLE 13.12

Orthotropic Thermal Conductivity Calculations for Digital PWB

Digital Board	CU Layer	Cu oz./mil	Type	thick (in.)	thick (m)	k, (W/m-K)	μ_i	Coverage		
								(k_{in}, top)	(k_{in}, bot)	(k_{out}, bot)
CU	1	1	Sig	0.0014	0.000036	390	0.3	0.00416052	0.00003556	3.03932E-07
Core		4	Core	0.004	0.000102	0.46	1	0.000046736	0.0001016	0.00022087
CU	2	1	GND	0.0014	0.000036	390	0.3	0.00416052	0.00003556	3.03932E-07
preg		4	preg	0.004	0.000102	0.46	1	0.000046736	0.0001016	0.00022087
CU	3	1	Sig	0.0014	0.000036	390	0.25	0.0034671	0.00003556	3.64718E-07
Core		5	Core	0.005	0.000127	0.46	1	0.00005842	0.000127	0.000276087
CU	4	1	Sig	0.0014	0.000036	390	0.3	0.00416052	0.00003556	3.03932E-07
preg		5	preg	0.005	0.000127	0.46	1	0.00005842	0.000127	0.000276087
CU	5	1	GND	0.0014	0.000036	390	0.3	0.00416052	0.00003556	3.03932E-07
Core		5	Core	0.005	0.000127	0.46	1	0.00005842	0.000127	0.000276087
CU	6	1	GND	0.0014	0.000036	390	0.3	0.00416052	0.00003556	3.03932E-07
preg		5	preg	0.005	0.000127	0.46	1	0.00005842	0.000127	0.000276087
CU	7	1	GND	0.0014	0.000036	390	0.9	0.01248156	0.00003556	1.01311E-07
Core		5	Core	0.005	0.000127	0.46	1	0.00005842	0.000127	0.000276087
CU	8	1	GND	0.0014	0.000036	390	0.9	0.01248156	0.00003556	1.01311E-07
preg		5	preg	0.005	0.000127	0.46	1	0.00005842	0.000127	0.000276087
CU	9	1	GND	0.0014	0.000036	390	0.9	0.01248156	0.00003556	1.01311E-07
Core		5	Core	0.005	0.000127	0.46	1	0.00005842	0.000127	0.000276087
CU	10	1	GND	0.0014	0.000036	390	0.9	0.01248156	0.00003556	1.01311E-07
preg		5	preg	0.005	0.000127	0.46	1	0.00005842	0.000127	0.000276087
CU	11	1	GND	0.0014	0.000036	390	0.9	0.01248156	0.00003556	1.01311E-07
Core		5	Core	0.005	0.000127	0.46	1	0.00005842	0.000127	0.000276087
CU	12	1	GND	0.0014	0.000036	390	0.3	0.00416052	0.00003556	3.03932E-07

#	layer	count	type							
	preg	5	preg	0.005	0.000127	0.46	1	0.00005842	0.000127	0.000276087
13	CU	1	Sig	0.0014	0.000036	390	0.3	0.00416052	0.00003556	3.03932E-07
	Core	5	Core	0.005	0.000127	0.46	1	0.00005842	0.000127	0.000276087
14	CU	1	Sig	0.0014	0.000036	390	0.25	0.0034671	0.00003556	3.64718E-07
	Core	4	Core	0.004	0.000102	0.46	1	0.000046736	0.0001016	0.00022087
15	CU	1	GND	0.0014	0.000036	390	0.3	0.00416052	0.00003556	3.03932E-07
	preg	4	preg	0.004	0.000102	0.46	1	0.000046736	0.0001016	0.00022087
16	CU	1	Sig	0.0014	0.000036	390	0.3	0.00416052	0.00003556	3.03932E-07
							Summation	0.107616244	0.00237236	0.003924406

copper	0.0224
dielectric	0.071
top+bot Solder	0
top+bot SM	0
TOTAL	0.0934

TABLE 13.13
Board Temperature Results of Select Devices on Slice PWB Assuming 70 °C Boundary

SWITCH Board Analog Parts

Location	Ref. Des.	Case Temp, °C	Theta_jc °C/W	Power, W	Junction Temp, °C
2A	Q111_26	96.96	0.5	0.105	92.18
5A FETs	Q11_30	96.9	0.5	0.656	94.94
5A Parts	R153	94.47	0	0.251	91.34
10A FETs	Q211_2	98.72	0.5	2.65	97.81
10A Res	R200	98.2	0	1.26	95.82
10A Parts	D106	99.5	300	0.019	102.3

TABLE 13.14
Board Temperature Results of Select Devices on Digital PWB Assuming 70 °C Boundary

Digital Board—PSS

Location	Case Temp, °C	Theta_jc °C/W	Power, W	Junction Temp, °C
U18	92.07	5	0.012	92.13
U300	76.12	5.6	0.34	78.02
U7	80.1	26	0.1	82.7
U1	82.65	26	0.17	88.2
U13	85.28	5.6	0.776	89.63
U2	85.89	2	0.685	87.26
L200	93.31	0	1.5	100.2
U250	94.71	8.3	1.75	108.2
L200	90.85	0	0.38	92.82
R303	93.23	0	0.15	93.23
Q230	91.98	1.67	2.5	93.3

14 Random Vibration Structural Analysis and Miles' Equation

INTRODUCTION

Random vibration studies are resource intensive and can easily be misconfigured, so a static representation is valuable for a quick-turn yet conservative alternative [1–4]. The goal is to determine peak response values. The resulting peak values may be used in a quasi-static analysis or perhaps in a fatigue calculation. The response levels could be used to analyze the stress in brackets and mounting hardware, for example.

LIMITING FACTORS

The resulting static load may be as high as one order of magnitude larger than the true dynamic load response of the system. One can use Miles' equation as a worst-case scenario whereby if it passes, then the design is rock-solid from a structural standpoint.

LOAD

These factors are critical in the structural analysis and must be agreed upon by all parties involved:

1. Mass, center-of-gravity, and inertia properties
2. Effective modal mass and participation factors
3. Stiffness
4. Damping
5. Natural frequencies
6. Dynamic mode shapes
7. Static deflection shape
8. Response acceleration
9. Modal velocity
10. Relative displacement
11. Transmitted force from the base to the component in each of three axes
12. Bending moment at the base interface about each of three axes
13. The manner in which the equivalent static loads and moments will be applied to the component, such as point load, body load, distributed load, and so on.
14. Dynamic stress and strain at critical locations if the component is best represented as a continuous system

DOI: 10.1201/9781003247005-14

15. Response limit criteria, such as yield stress, ultimate stress, fatigue, and loss of clearance

Each of the response parameters should be given in terms of frequency response function, power spectral density, and overall response level.

Furthermore, assumptions must be documented, including a discussion of conservatism.

MODAL RESPONSE

Of the three motion parameters (displacement, velocity, and acceleration) describing a shock spectrum, velocity is the parameter of greatest interest from the viewpoint of damage potential. This is because the maximum stresses in a structure subjected to a dynamic load typically are due to the responses of the normal modes of the structure, that is, the responses at natural frequencies. At any given natural frequency, stress is proportional to the modal (relative) response velocity. Specifically:

$$\sigma_{max} = C V_{max} \sqrt{E \rho}$$

(14.1)

where

σ_{max} = Maximum modal stress in the structure

V_{max} = Maximum modal velocity of the structural response

E = Elastic modulus

ρ = Mass density of the structural material

C = Constant of proportionality dependent upon the geometry of the structure (often assumed for complex equipment to be $4 < C < 8$)

RELATIVE DISPLACEMENT

Note that the transmitted force for a single-degree-of-freedom (SDOF) system is simply the mass times the response acceleration. Specifying the relative displacement for an SDOF system may seem redundant because the relative displacement can be calculated from the response acceleration and the natural frequency per equation (14.7) given later in this chapter.

The reason is that the relationship between the relative displacement and the response acceleration for a multi-degree-of-freedom (MDOF) or continuous system is complex. Any offset of the component's center of gravity (CG) further complicates the calculation due to coupling between translational and rotational motion in the modal responses.

MODEL

Determination of the acceleration response of the component requires modelling the component as an SDOF system, if appropriate, as shown in Figure 14.1.

FIGURE 14.1 Single-degree-of-freedom system.

where

m   is the mass
C   is the viscous damping coefficient
k   is the stiffness
x is the absolute displacement of the mass
y is the base input displacement

Furthermore, the relative displacement z is:

$$z = x - y$$

(14.2)

The natural frequency of the system fn is:

$$fn = \frac{1}{2\pi}\sqrt{\frac{k}{m}}$$

(14.3)

ACCELERATION RESPONSE

Miles' equation is a simplified method of calculating the response of an SDOF system to a random vibration base input, where the input is in the form of a power spectral density.

The overall acceleration response $\ddot{x}_{GRMS}$ is:

$$\ddot{x}_{GRMS}\left(f_n, \xi\right) = \sqrt{\left(\frac{\pi}{2}\right)\left(\frac{fn}{2\xi}\right)P}$$

(14.4)

where

TABLE 14.1
Miles' Equation References

Reference	Author	Equation	Page
1	Himelblau	(10.3)	246
2	Fackler	(4–7)	76
3	Steinberg XE "Steinberg"	(8–36)	225
4	Luhrs	-	59
5	Mil-Std-810G	-	516.6–12
6	Caruso	(1)	28

Fn   is the natural frequency
P   is the base input acceleration power spectral density at the natural frequency
ξ is the damping ratio

Damping is represented in terms of the quality factor Q:

$$Q = \frac{1}{2\xi}$$

$$(14.5)$$

Equation (14.4), or an equivalent form, is given in numerous references, including those listed in Table 14.1.

Furthermore, Miles' equation is an approximate formula that assumes a flat power spectral density from zero to infinity Hz. As a rule of thumb, it may be used if the power spectral density is flat over at least two octaves centered at the natural frequency.

RELATIVE DISPLACEMENT AND SPRING FORCE

Consider an SDOF system subject to a white noise base input and with constant damping. The Miles' equation set shows the following with respect to the natural frequency fn:

Response Acceleration	$\propto \sqrt{f_n}$	(14.6)
Relative Displacement	$\propto 1 / f_n^{1.5}$	(14.7)
Relative Displacement	$= \text{Response Acceleration} \ / \ \omega_n^2$	(14.8)

where

$$\omega_n = 2\pi f_n$$

Consider that the stress is proportional to the force transmitted through the mounting spring. The spring force F is equal to the stiffness k times the relative displacement z:

$$F = kz \tag{14.9}$$

RMS AND STANDARD DEVIATION

The root mean square (RMS) value is related the mean and standard deviation σ values as follows:

$$\text{RMS}^2 = \text{mean}^2 + \sigma^2 \tag{14.10}$$

Note that the RMS value is equal to the 1σ value assuming a zero mean.
A 3σ value is thus three times the RMS value for a zero mean.

REFERENCES

1. Irvine, Tom, *Equivalent Static Loads for Random Vibration*, Revision M, October 8, 2010.
2. James, M.L., G. Smith, J., Wolford, and P. Whaley, *Vibration of Mechanical and Structural Systems*, Harper & Row, New York, 1989.
3. Steinberg, Dave, *Vibration Analysis for Electronic Equipment*, 3rd Ed., Wiley InterScience, Danvers, MA, 2000.
4. Jamnia, Ali, *Practical Guide to the Packaging of Electronics*, 3rd Ed., CRC, Boca Raton, FL, 2016.

15 Vibrational Analysis Case Studies

INTRODUCTION TO THE VIBRATIONAL CASE STUDIES

The goal of a vibrational analysis is to increase assurance that the hardware will be able to handle mainly the vibrational environment caused by the launch vehicle rocket. Vibration tends to be the harshest of the structural environments, followed by pyrotechnic and impact shock. What makes vibration concerning during launch are the repetitive cycles that the printed circuit board will experience, which will translate into cyclic fatigue stresses on the connections and solder joints at the modal frequencies. The hardware must also be able to endure simulated vibration cycles during the vibration acceptance testing, and in the three major seconds.

All the mechanical screws associated with the chassis enclosure must also be analyzed to show margin against slippage and gapping, as sometimes designs with too few screws will tend to fall apart during testing and possibly subject flight hardware to damage. The chassis walls of the enclosure need analysis as well, as the chassis may be subjected to too aggressive light-weighting to reduce overall mass. Finally, small electronic parts with short, incompliant connection leads can sometimes be sensitive to shock waves that can be caused by pyrotechnics employed during launch, for example, or during fairing separation. All this leads to catching weak points in the design that can be fixed before any metal is cut.

INTEROFFICE MEMORANDUM
Project: Mars 2020
November 12, 2018

SUBJECT: PHOTODIODE BOARD ASSEMBLY STRUCTURAL ANALYSIS

REF: 1. GEVS GFSC-STD-7000A
  2. *Vibration Analysis for Electronic Equipment*, 3rd Edition, by Dave Steinberg

CONCLUSION

A structural analysis was performed for the photodiode for the laser power supply (LPS) aboard SHERLOC on Mars 2020. The maximum expected board displacement is 0.003423 inch against a maximum allowable of 0.0249 inch to satisfy 20 M fatigue cycles resulting in a factor of safety of 7.29. There is no concern for fatigue failure.

DOI: 10.1201/9781003247005-15

MODEL ASSUMPTIONS

A structural model of the complete photodiode slice assembly including laser conditioning optics (Figure 15.1) and mounting plate was created using SOLIDWORKS Simulation finite element code and composed of 22,363 nodes and 11,652 elements (Figure 15.2). The photodiode board assembly mass is 2 grams.

A structural analysis of the photodiode slice assembly was performed using GEVS 6.8 G_{rms} random vibration spectrum. The spectrum is applied in the worst-case direction with damping at 2.5%. The PWA deflection response was found to be within the guidelines of (Steinberg) [1] for a vibration fatigue life of 20 million cycles, which is 8 hours of vibration at the fundamental PWB vibration frequency of 573.4 Hz (see Figure 15.3a); the second mode is shown in Figure 15.3b. Total model mass is 16.7 grams, of which approximately 2 grams was of the photodiode board assembly.

A random vibration analysis showed the first three resonant frequency modes, Figure 15.3d shows eighth mode, which was shown to have the largest mass participation (Table 15.1) at 3093 Hz, which is below the definition of the PSD curve shown in Figure 15.4 and Table 15.1 and believed to have an insignificant stress response, and is therefore not included in the analysis. The first three modes were analyzed with Miles' equation shown in Table 15.2, and the first mode was shown to have the largest response and was used in the static analysis.

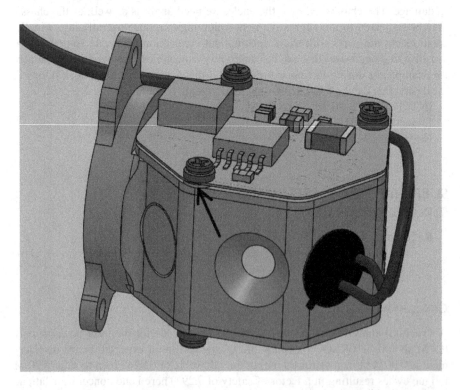

FIGURE 15.1 Photodiode board assembly mounted on top of the laser conditioning optics.

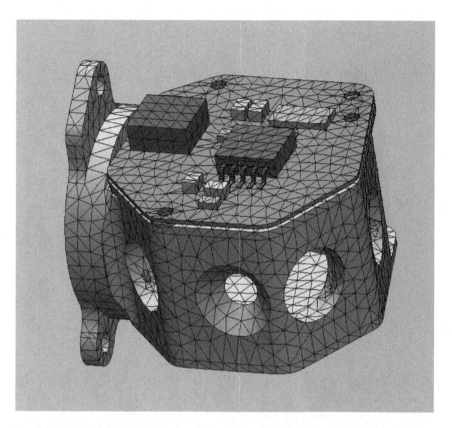

FIGURE 15.2 The photodiode board assembly meshed in SOLIDWORKS Simulation.

RESULTS

For the given qualification vibration spectrum density of 0.035 g^2/Hz at 573 Hz, the expected 3 sigma board displacement is 0.003423 inch as shown in Figure 15.6.

Using the Steinberg method shown in Table 15.3 for 20 million cycles of fatigue endurance, the board should not deflect more than 0.0053 inch. Therefore, there is an expected factor of safety of .00249 / .003423 = 7.29.

The structural analysis shows there is no concern about fatigue failure.

INTEROFFICE MEMORANDUM

Project: Sentinel-6

November 8, 2017

SUBJECT: SINE VIBRATION ANALYSIS OF HRMR DDU FOR SENTINEL-6

REF: 1. GEVS GFSC-STD-7000A

          2. *Vibration Analysis for Electronic Equipment*, 3rd Edition, by Dave Steinberg

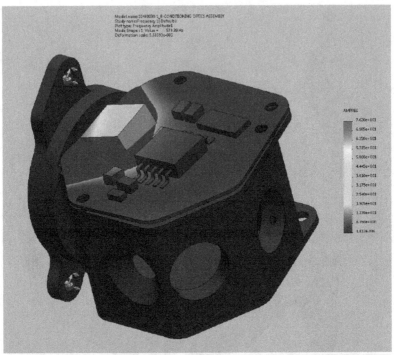

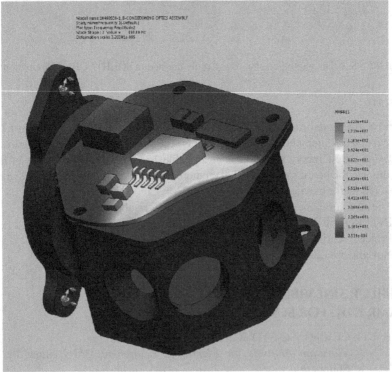

FIGURE 15.3 (a) First mode at 573.4 Hz; (b) second mode at 898 Hz.

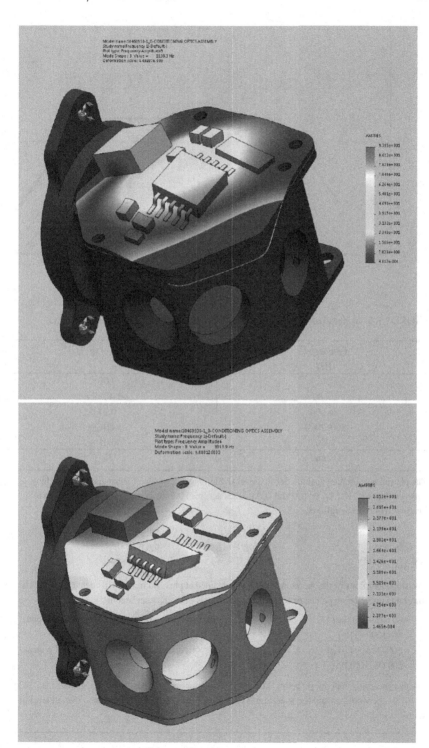

FIGURE 15.3 (c) Third mode at 1138 Hz; (d) eighth mode with largest mass participation.

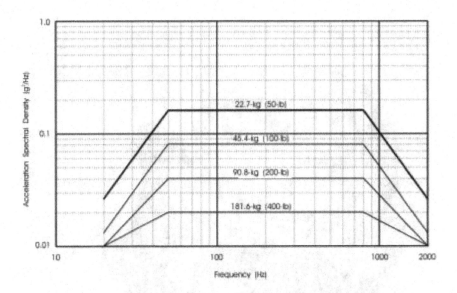

FIGURE 15.4 Generalized random vibration test levels for 22.7kg (50 lb.) or less.

Frequency (Hz)	ASD Level (g²/Hz)
20	0.01
20–80	+3 dB/oct
80–500	0.04
500–2000	−3 dB/oct
2000	0.01
Overall	$6.8 \, G_{rms}$

The plateau acceleration spectral density level (ASD) may be reduced for components weighing between 45.4 and 182 kg, or 100 and 400 pounds according to the component weight (W) up to a maximum of 6 dB as follows:

	Weight in kg	Weight in lb
dB reduction	$= 10 \log(W/45.4)$	$10 \log(W/100)$
$ASD_{(plateau)}$ level	$= 0.04 \cdot (45.4/W)$	$0.04 \cdot (100/W)$

The sloped portions of the spectrum shall be maintained at plus and minus 3 dB/oct. Therefore, the lower and upper break points, or frequencies at the ends of the plateau become:

$F_L = 80 \, (45.4/W)$ [kg] F_L = frequency break point low end of plateau
  $= 80 \, (100/W)$ [lb]

$F_H = 500 \, (W/45.4)$ [kg] F_H = frequency break point high end of plateau
  $= 500 \, (W/100)$ [lb]

The test spectrum shall not go below 0.01 g²/Hz. For components whose weight is greater than 182 kg or 400 lb, the workmanship test spectrum is 0.01 g²/Hz from 20 to 2000 Hz with an overall level of 4.4 G_{rms}.

FIGURE 15.5 Spectral density curve from GEVS Table 15.2 for 100 lb or less.

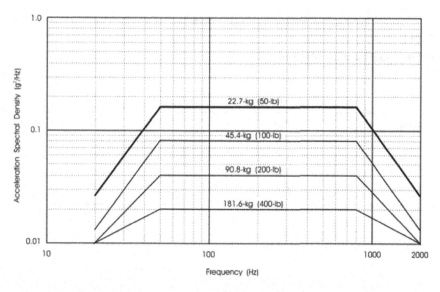

FIGURE 15.6 Spectral density values used in creation of PSD curve in Figure 15.4.

TABLE 15.1
Mass Participation List

Mode No.	Freq (Hertz)	x-direction	y-direction	z-direction
1	573.39	0.020195	0.010764	0.0018847
2	898.08	0.01613	0.0085264	0.00018556
3	1138.3	0.0085645	0.0030826	0.00068579
4	1956.4	0.0014287	0.00040419	3.67E-05
5	2051.4	0.0095376	0.0022351	0.00043321
6	2392.5	1.93E-05	9.79E-05	0.00094801
7	2817.4	0.0045233	0.00025925	0.00077421
8	3093.9	0.14173	0.39468	0.0019427
9	3396.4	0.01842	5.52E-05	0.0035913
10	3487.8	0.080305	0.019377	0.057594
11	3905.7	0.21448	0.102	0.041704
12	3990.8	0.012501	0.0016305	0.00017231
13	4201	0.0060674	0.0056627	0.0015381
14	4307.9	0.0054478	0.00038881	0.00067923
15	5267	0.0027949	0.00034308	0.0015877
16	5854.7	0.015062	0.0026513	0.014418
17	6156.4	0.0096283	0.0030618	0.019063
18	6747.6	0.0001725	3.79E-05	0.001679
19	6847.8	0.00027994	0.00036332	0.00047808
20	7640.7	0.00020749	0.0011753	1.34E-05

TABLE 15.2
Miles' Equation Calculations

Program	$\ddot{\chi}_{G_{rms}}$	$\dfrac{G^2}{HZ}$ PDS	HZ fn	$\dfrac{1}{2\xi}$ Q	3σ	G, m/s^2
1st mode	25.11	0.035	573	20	75.32	738.97
2nd mode	24.61	0.023	838	20	73.82	724.19
3rd mode	24.57	0.017	1130	20	73.7	722.98

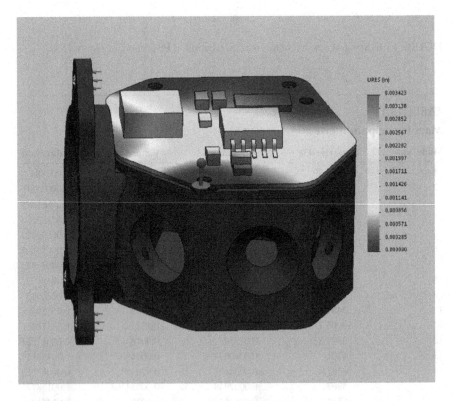

FIGURE 15.7 Results of random vibration static analysis showing a maximum board deflection of 0.003423 inches at first mode resonant frequency.

TABLE 15.3
Steinberg Deflection Analysis Coefficients

Max PCB deflection 3σ

Z, in.	Z, mm	B	L	h	C	r	P
0.0249	0.634	1.2	0.28	0.02	1	1	0.035

Modeled 3s deflection, in.
0.0034

Factor of Safety	**7.29**

TABLE 15.4
ESA Sine Vibration Environment

Components	Freq (Hz)	Acceleration (G_{rms})	Protoflight
ESA Panel Components:	5	3.2	4
CMIE, DDU, EUs	20	16.8	21
	100	16.8	21
Duration:	2 oct/min		Qual: 2 oct/min PF: 4 oct/min

The digital data unit (DDU) assembly design was found acceptable for random vibration. Additionally, a sine vibration was performed for the environment in Table 15.4 provided by ESA.

A linear dynamic structural analysis of the DDU assembly was performed using Qual Environment 21 G_{rms} sine vibration applied between 5 Hz and 100 Hz. The modal vibration frequencies listed are all excited by vibration in the axis perpendicular to the mounting plane. The vibration amplification response of the PWB at 100 Hz was found to be .00066", which was close to the static acceleration of 21g .00065" at 100 Hz. This is small compared to the allowable .0192" calculated in the previous random vibration structural interoffice memo (IOM). A sine vibration test is believed to be benign compared to the random vibration test.

DDR modal frequency, Hz mode 1 is the cover and mode 2 is the PWA.

Deflection response of the PWB versus sine frequency is shown in Figure 15.10. The 5–100 Hz sine vibration spectrum was applied in 10 steps starting at 5 Hz. The deflection was measured at a node near the maximum deflection node, which is why the largest value is slightly less than the reported maximum of .00066".

INTEROFFICE MEMORANDUM
Project: Mars 2020
April 05, 2019

TABLE 15.5

Mode and Frequency Results From Sine Vibration Environment

Mode No.	Freq (Hertz)
1	581.6
2	660.58
3	729.74
4	732.84
5	761.29
6	813.11
7	833.43
8	976.09
9	1245.7
10	1274
11	1389.7
12	1510
13	1557.2
14	1613.9
15	1691.3

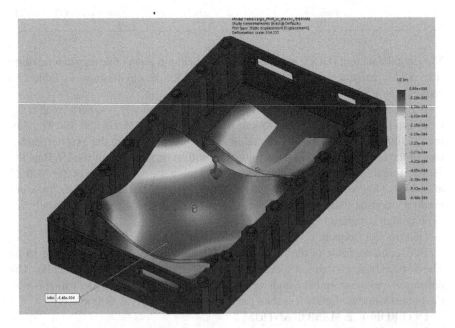

FIGURE 15.8 Deflection caused by static acceleration of 21 g, measured to be .00065″. The deflection is exaggerated in the image.

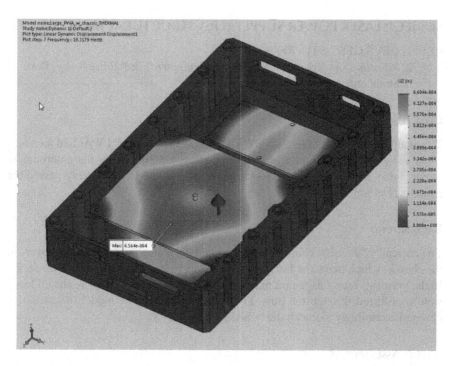

FIGURE 15.9 Deflection caused by 21 g sine vibration at 100 Hz, measured to be .00066".

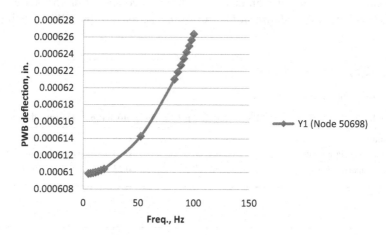

FIGURE 15.10 PWB deflection as a function of frequency.

SUBJECT: LASER POWER SUPPLY (LPS) STRUCTURAL ANALYSIS

REF: 1. GEVS GFSC-STD-7000A

      2. *Vibration Analysis for Electronic Equipment*, 3rd Edition, by Dave Steinberg

CONCLUSION

The analysis showed that the main PWA as well as the cap bank PWAs had no issues when subjected the random vibration environment stipulated by the environment requirements document (ERD). A venting analysis of the LPS assembly passed the guidelines as well.

BACKGROUND

The current "c" revision contains changes to the fastening approach of the covers to the chassis, which treats the fasteners more like a bolted joint across a surface area. In the prior two revisions, 2 mm pins were put into the screw locations and all loads were transferred through the pins. The fixing scheme of the model in space also changed accordingly to match the bolted-joint geometry.

MODEL ASSUMPTIONS

A structural model of the complete laser power supply chassis assembly, including both main PWA and capacitor banks PWAs, was created using SOLIDWORKS Simulation finite element high-quality tet-10 mesh composed of 153,384 elements and 283,667 nodes, as shown in Figure 15.10. The mass of the LPS assembly was

TABLE 15.6
Material Properties

Material	Elastic Modulus, psi	Poisson's Ratio	Density, lbs./in³
Aluminum	Aluminum	0.33	0.1
Polyimide XE "Polyimide" -glass	3,500,000	0.15	as required for PWA mass
BaTiO3 (capacitor bank)	9,700,000	0.23	0.2

TABLE 15.7
Mechanical Properties of Enclosure Materials

Material	Ultimate Tensile Strength	Yield Strength
Al 7075-T73	500e+06 N/m^2 (500 MPa)	500e+06 N/m^2 (410 MPa)
A286 Fasteners	620e+06 N/m^2 (620 MPa)	275e+06 N/m^2 (275 MPa)

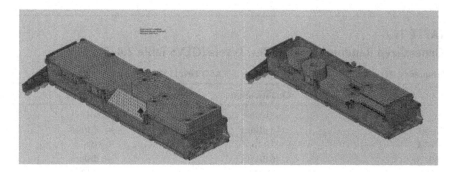

FIGURE 15.11 FEM mesh of the structural model. The green arrows show how the model is fixed in space.

FIGURE 15.12 The fastener locations are modeled by attachment between a 4 mm diameter annulus split plate for the LPS and an annulus that is 2x the hole diameter on the bracket and flexure.

660 grams. A structural analysis of the LPS assembly was performed using the ERD Qual Environment 7.9 G_{rms} random vibration spectrum (see Figure 15.10). The spectrum was applied in the worst-case direction with damping at 2%.

The main PWB was attached to the chassis using 11 perimeter M2 screws on the top cover and 11 perimeter screws on the bottom cover. The PWB is wet-wet-mounted to the chassis with Nusil 2946 thermal bonding adhesive around the entire perimeter. The chassis attachment to the turret mechanical structure is through a bracket and a flexure at each end (shown by the green arrows in Figure 15.10.

MODELING OF THE FASTENERS

The bolted joint on the covers to the chassis are simulated by using 4 mm split planes at each fastener location, including the fasteners where the LPS is fastened to the rest of the rover turret assembly (see Figure 15.11). The rest of the mating surfaces of the

TABLE 15.8

Generalized Random Vibration Test Levels (GEVS Table 2.4–3)

Frequency (Hz)	ASD Level (g2/HZ)	
	Qualification	Acceptance
20	0.026	0.013
20–50	+6 dB/oct	+6 dB/oct
50–800	0.16	0.08
800–2000	−6 dB/oct	−6 dB/oct
2000	0.026	0.013
Overall	14.1 G_{rms}	10.0 G_{rms}

The acceleration spectral density level may be reduced for components weighing more than 22.7 kg (50 lb) according to:

	Weight in kg	Weight in lb	
dB reduction	= 10 log(W/22.7)	10 log(W/50)	
$ASD_{(50-800\ Hz)}$	= 0.16·(22.7/W)	0.16·(50/W)	for protoflight
$ASD_{(50-800\ Hz)}$	= 0.08·(22.7/W)	0.08·(5Q/W)	for acceptance

Where W = component weight.

The slopes shall be maintained at + and −6dB/oct for components weighing up to 59 kg (130 lb). Above that weight, the slopes shall be adjusted to maintain an ASD level of 0.01 g²/Hz at 20 Hz and 2000 Hz.

For components weighing over 182 kg (400 lb), the test specification will be maintained at the level for 182 kg (400 lb).

top and bottom covers to the chassis are given contact surface parameter "allow to penetrate," which simulates a gap. The LPS flexure and bracket 3 screw locations, three for each side, are also modeled as split plates annuli that are 2x the hole diameter, and they are fixed in space by fixing their split planes in space.

RANDOM VIBRATION ENVIRONMENT

The random vibration linear dynamic analysis was performed within SOLIDWORKS Simulation using linear dynamic with a random vibration input. Figure 15.11 shows the 0.08 g^2/Hz PSD acceleration that was used in conjunction with the frequency curve shown used to represent the random vibration environment. The same 0.08 g^2/Hz environment was applied in all three axes and the maximum stress values for the chassis, covers, and screws were reported, as well as the maximum deflection values for the PWBs. Mass participations and frequency response across all three axes are shown in Figures 15.12, 15.13, and 15.14.

Z-AXIS RESPONSE

The response graph in the z-direction is shown below in Figure 15.13 with the primary mode occurring at 414 Hz.

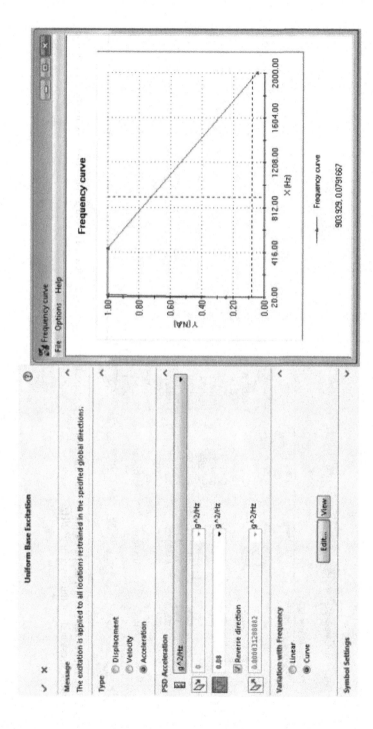

FIGURE 15.13 Random vibration base excitation parameters and frequency curve used to represent the GEVS 100 lb vibration environment.

Mass Participation (Normalized)

Study name:dynamic-Z

Mode No.	Freq [Hertz]	X direction	Y direction	Z direction
1	413.67	0.00015281	0.066983	0.70935
2	684.16	0.00020945	0.48857	0.083185
3	889.39	0.00054	0.028136	8.3149e-008
4	947.53	0.0175	0.0031352	0.00055696
5	1154.3	0.044247	0.00015626	0.0027795
6	1168.3	0.087756	0.052347	0.02724
7	1193.9	0.23633	0.024503	0.018431
8	1270.5	0.44748	0.0089128	0.045938
9	1429.8	0.0090372	0.020107	0.0032967
10	1466.1	0.00058026	0.021941	0.0037529
11	1537	0.0090325	0.047269	0.0060239
12	1601.6	0.0016829	0.0050269	0.0025776
13	1675.7	0.03548	0.091419	0.00017389
14	1836.4	0.0010205	8.0716e-005	8.0968e-007
15	1898.5	0.032164	0.00055614	0.0046771
		Sum X = 0.92319	Sum Y = 0.87914	Sum Z = 0.90798

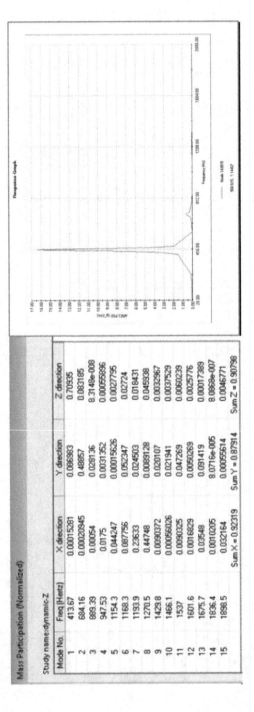

FIGURE 15.14 The z-axis mass participations and modes at their respective resonant frequencies. The red graph represents the response of the maximum stress node of the cover in g^2/Hz.

Y-Axis Response

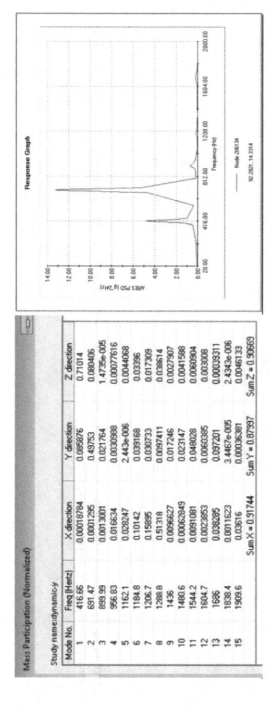

Mass Participation (Normalized)

Study name:dynamic-y

Mode No.	Freq [Hertz]	X direction	Y direction	Z direction
1	416.66	0.00018784	0.095876	0.71014
2	691.47	0.0001295	0.49753	0.080406
3	899.99	0.0013001	0.021764	1.4735e-005
4	956.83	0.016634	0.0030988	0.00077616
5	1162.1	0.028247	2.443e-006	0.0044068
6	1184.8	0.10142	0.039168	0.03396
7	1206.7	0.15895	0.030733	0.017309
8	1288.8	0.51318	0.0097411	0.038614
9	1436	0.0096627	0.017246	0.0027907
10	1480.6	0.00062849	0.023147	0.0041568
11	1544.2	0.0091081	0.048028	0.0060904
12	1604.7	0.0023853	0.0060385	0.003008
13	1686	0.038285	0.097201	0.00039311
14	1838.4	0.0011623	3.4467e-005	2.4343e-006
15	1909.6	0.03616	0.00036381	0.0046133
		Sum X = 0.91744	Sum Y = 0.87397	Sum Z = 0.90669

FIGURE 15.15 The *y*-axis mass participations and modes at their respective resonant frequencies. The red graph represents the response of the maximum stress node of the cover in g^2/Hz.

X-Axis Response

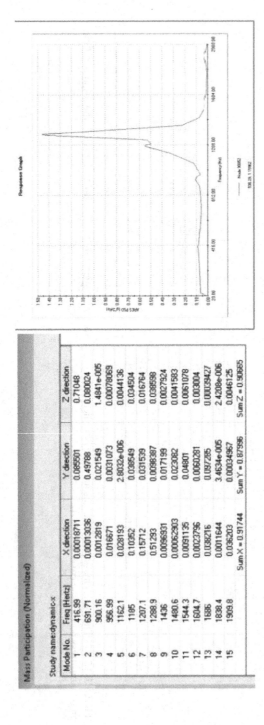

Mass Participation (Normalized)

Study name:dynamic-x

Mode No.	Freq (Hertz)	X direction	Y direction	Z direction
1	416.99	0.00018711	0.065501	0.71048
2	691.71	0.00013036	0.49788	0.060024
3	900.16	0.0012819	0.021549	1.4841e-005
4	956.99	0.016671	0.0031073	0.00078069
5	1162.1	0.028193	2.8032e-006	0.0044136
6	1185	0.10352	0.038549	0.034504
7	1207.1	0.15712	0.031539	0.016764
8	1288.9	0.51293	0.0098387	0.038598
9	1436	0.0096931	0.017199	0.0027924
10	1480.6	0.00062903	0.023082	0.0041583
11	1544.3	0.0091135	0.04801	0.0061078
12	1604.7	0.00023796	0.0060281	0.003004
13	1686	0.038216	0.09/265	0.00039427
14	1838.4	0.0011644	3.4634e-005	2.4209e-006
15	1909.8	0.036203	0.00034567	0.0046125
		Sum X = 0.91744	Sum Y = 0.87996	Sum Z = 0.90685

FIGURE 15.16 The z-axis mass participations and modes at their respective resonant frequencies. The red graph represents the response of the maximum stress node of the cover in g^2/Hz.

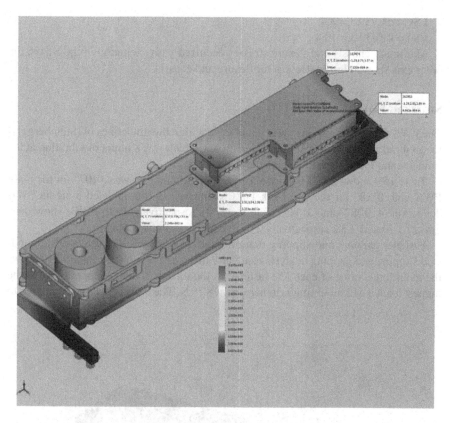

FIGURE 15.17 Primary mode of cap bank PWB and main PWA was 413 Hz and the RMS board displacements are shown.

RESULTS: MAXIMUM DEFLECTION OF THE PWAS

Figure 15.5 shows the maximum RMS deflection of the main PWA, and the cap bank PWA, in the out of plane (z-axis) relative to the PWBs. The measure board displacements were as followed and used in calculating fatigue life estimates discussed later in the report.

1σ Cap bank = .0033 — .00041" = .0029",
1 σ σ Main PWB = .0022 — .0007 = .0015"
3 σ Cap bank = .0087"
3 σ Main PWB = .0045"

RANDOM VIBRATION RESULTS: CHASSIS STRESSES

The maximum 3-sigma von Mises stress on the chassis was 39.3 MPa as shown in Figure 15.17, which resulted in a factor of safety for yield and ultimate tensile strength for AL 7071-T73 as follows:

3s von Mises = 3x(13.1 MPa) = 39.3 MPa

FSy = 330/39.3 = 8.4
FSu = 410/39.3 = 10.4

Maximum chassis and flexure stresses occurred with excitation in the z-axis; the other axis showed lesser stresses for flexure and chassis.

STEINBERG FATIGUE ANALYSIS [1]

The PWA deflection response was found to be within the guidelines of (Steinberg) [1] for a vibration fatigue life of 20 million cycles, which was 8 hours of vibration at the fundamental PWB vibration frequency of 414 Hz.

Table 15.9 shows PWB flexure allowance per (Steinberg) was .0107" for the main PWA and .0148" for the cap bank PWB compared to the modeled 3-sigma PWB deflection of .0087" and .0045" (Figure 15.14), respectively. Both PWBs show margins of 1.70 and 2.38, respectively. Positive margins are acceptable.

Fastener gapping and slipping margin Table 15.10 shows the results of the modeled slipping and gapping analysis results for the chassis fasteners. Table 15.10 shows the axial force value of each M2 fastener in the model. 3-sigma gapping load at the highest loaded M2 mounting fastener was 118.75 N. The margin is large compared to

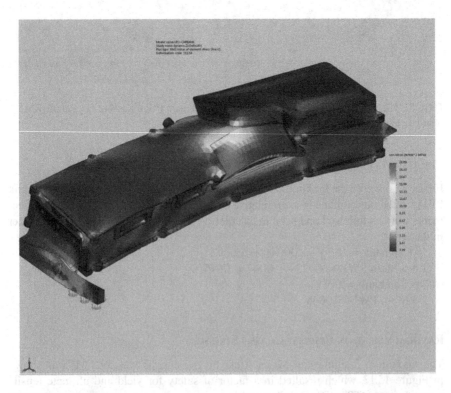

FIGURE 15.18 Maximum chassis and flexure RMS von Mises stress: 13.09 MPa and 9.61 MPa, respectively.

TABLE 15.9
Steinberg Fatigue Analysis Results

	1.26	Steinberg PCB Deflection Analysis					
Device	Max PCB deflection 3-sigma, inch		PCB Length, in	Compt length	PCB t	Constant	Placement
	Z, in.	Z, mm	B	L	h	C	r
LPS Main M3 & M	0.0107	0.272	9.100	0.787	0.094	2.250	1.000
LPS Caps PWBs	0.0148	0.376	9.843	1.000	0.098	1.500	1.000

Modeled 3-Sigma Deflection—Cap Bank		Modeled 3-Sigma Deflection—Main PWA	
0.0087		0.0045	
Margin	1.70	Margin	**2.38**

the expected M2 torqued reduced preload. 3-sigma slipping load was 40.93 N. The margin was large compared to the slipping capability of 168.75 N, or FS of +9.67. Positive margins are acceptable.

There are six screws that make up the reaction attachment points, three on each side of the module (shown in Table 15.11). The maximum axial stresses were seen in the z-direction. The slipping capability margin was 0.61; positive margins are acceptable.

VENTING ANALYSIS

A venting analysis was conducted on the LPS mechanical model. The LPS was measured to have a total of 3.5 E+04 m^3 of air space inside the module. The space above and below the main PWA were assumed to be one space due to the generous amount of feedthrough area for wire routing from top to bottom of the main PWA, well below the necessary V/A requirement, where V is volume, and A is the feedthrough area. A pressure of 0.6 psi was applied normal to all the flat inner surfaces of the chassis and the two lids, causing a "ballooning" effect on the inner chassis and covers. This was the assumed pressure differential based on a standard guidelines. The result shown in Figure 15.18 was a maximum stress at the chassis of 13.66 MPa for an ultimate tensile strength of 620 MPa for Al 7075-T73, renders a factor of safety of 20.1, meeting the recommended factor safety of 2.0.

Maximum von Mises stress = 13.66 MPa

FSy = 330/13.66 = 24.2
FSu = 620/13.66 = 30.0

TABLE 15.10

Gapping and Slipping Factor of Safety Results

Fastener	Axial Force (N)	Shear1(N)	Shear2(N)	3s Gapping Load (N)	3s Slipping Load (N)	Reduced Load (N)	Slipping Capability (N)	Red. Load/ Gap. Load	Slip Cap/ Slip Load
MAX	37.70	4.78	3.39	113.10	17.57	849.40	169.88	7.51	9.67
MJ2x.7					N				

Max Preload Table 14 JPL D-51878 = 1250 N (282.81 lbf)
Minimum Preload: 77% of Original = 962.5
Coefficient of friction—mating assemblies = 0.2

M2 Area	3.14159E-06	m^2
1-sigma axial stress	1.20E+07	N/m^2
1-sigma shear stress 1	1.52E+06	N/m^2
1-sigma shear stress2	1.08E+06	N/m^2

TABLE 15.11

Gapping and Slipping Forces of the Reaction Screws

Reaction Screws M2.5

Fastener	Axial Force (N)	Shear 1(N)	Shear 2(N)	3s Gapping Load (N)	3s Slipping Load (N)	Reduced Load (N)	Slipping Capability (N)	Red. Load/ Gap. Load	Slip Cap/Slip Load
MAX	106.94	29.04	63.33	320.83	209.03	641.67	128.33	2.00	0.61
MJ2x.7	Max Preload Table 14 JPL D-51878 = 1250 N (282.81 lbf)				N				
	Minimum Preload: 77% of Original = 962.5								
	Coefficient of friction—mating assemblies = 0.2								
	M2.5 Area	4.52389E-06	m^2						
1-sigma	axial stress	2.36E+07	N/m^2						
1-sigma	shear stress 1	6.42E+06	N/m^2						
1-sigma	shear stress2	1.40E+07	N/m^2						

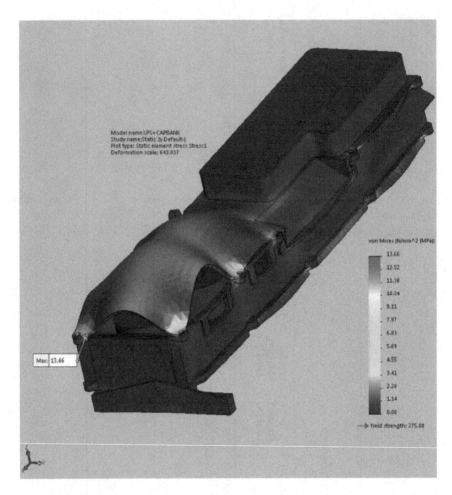

FIGURE 15.19 von Mises maximum stress associated with venting analysis showing maximum stress of 13.66 MPa at the top cover.

Maximum von Mises and Shear (TauXY) Stresses Across All M2 Fasteners

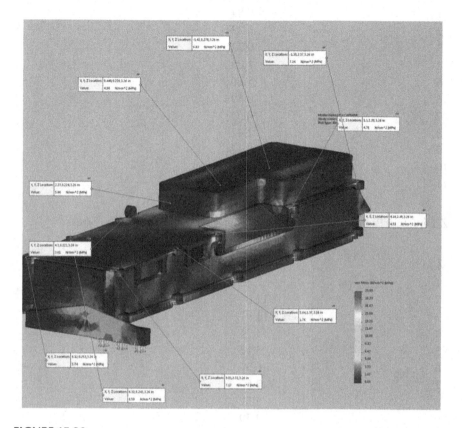

FIGURE 15.20

Z-Direction Top Fasteners Normal (von Mises) and Shear (Tau) Stress

M2	Stress, MPa
1	3.61
2	5.86
3	4.98
4	6.93
5	7.24
6	4.70
7	4.53
8	1.76
9	7.17
10	8.50
11	5.74

M2	TauXY, MPa
1	0.84
2	0.23
3	0.22
4	1.03
5	1.06
6	0.89
7	0.54
8	1.42
9	0.12
10	3.52
11	0.66

Z-Directoin Bottom Fasteners Normal (von Mises) and Shear (Tau) Stress

M2	Stress, MPa
1	7.36
2	12.0
3	11.97
4	4.34
5	3.12
6	6.37
7	7.53
8	5.57
9	2.59
10	3.64
11	4.0

M2	TauXY, MPa
1	0.41
2	1.10
3	1.13
4	0.23
5	0.39
6	0.09
7	0.31
8	0.19
9	0.27
10	0.22
11	0.48

Y-Direction Top Fasteners Normal (von Mises) and Shear (Tau) Stress

M2	vMises, MPa
1	2.83
2	3.0
3	1.63
4	3.95
5	2.74
6	8.94
7	2.06
8	1.95
9	1.36
10	2.34
11	1.83

M2	TauXY, MPa
1	0.08
2	0.15
3	0.33
4	1.52
5	0.52
6	1.31
7	0.78
8	0.15
9	0.13
10	1.12
11	0.74

Y-Directoin Bottom Fasteners Normal (von Mises) and Shear (Tau) Stress

M2	vMises, MPa
1	1.91
2	3.01
3	3.96
4	0.87
5	2.08
6	2.30
7	2.39
8	2.48

M2	vMises, MPa
9	1.73
10	0.81
11	1.99

M2	TauXY, MPa
1	0.24
2	0.38
3	0.41
4	0.27
5	0.14
6	0.16
7	0.08
8	0.44
9	0.11
10	0.25
11	0.16

X-Direction Top Fasteners Normal (von Mises) and Shear (Tau) Stress

M2	vMises, MPa
1	0.34
2	0.14
3	0.76
4	2.62
5	1.5
6	1.4
7	0.49
8	0.29
9	0.53
10	0.81
11	0.32

M2	TauXY, MPa
1	0.12
2	0.017
3	0.076
4	0.250
5	1.07
6	0.40

M2	TauXY, MPa
7	0.70
8	0.11
9	0.11
10	0.05
11	0.41

X-Direction Bottom Fasteners Normal (von Mises) and Shear (Tau) Stress

M2	vMises, MPa
1	0.24
2	0.40
3	0.90
4	0.26
5	0.58
6	1.51
7	1.21
8	0.50
9	0.26
10	0.19
11	0.39

M2	TauXY, MPa
1	0.04
2	0.05
3	0.05
4	0.07
5	0.06
6	0.05
7	0.06
8	0.20
9	0.02
10	0.06
11	0.03

REFERENCE

1. Steinberg, D., *Vibration Analysis for Electronic Equipment*, 3rd Ed., John Wiley & Son, Boca Raton, FL, 2000.

16 Creep Prediction of a Printed Wiring Board for Separable Land Grid Array Connector

INTRODUCTION

High pin count and high-density electronics continue to push electronic packaging materials to their limits. The next generation of high-performance packages such as multichip modules (MCM), flip-chip, chip scale ball grid array (BGA), and so on requires low-cost, high-density and high-reliability PWB for high-density interconnect (HDI) [1]. Larger but more densely packed second- and third-level packages require larger temperature stability across large spans. An alternative to solder ball and solder columns on second-level packages are land grid array (LGA) connectors that achieve connection by compression of compliant pins, which have their set of advantages and disadvantages. A major disadvantage is the requirement of board flatness over time that may require bulky and heavy mechanical hardware to achieve. Another approach is to use the high stiffness and temperature stability of the PWB material to help in achieving these requirements. A class of PWB materials is available with high stiffness and high temp stability such as the classic FR4, high-temperature FR4, GETEK, and NELCO materials, which are all essentially epoxy/glass polymeric matrix composites. The material of choice of this experiment is GETEK. The concern addressed here is relaxation over time, or creep of the material, because it is a general understanding that all polymeric materials will creep. The question is how much they will creep under a known set of conditions and whether this creep is within an acceptable range. It is understood that woven fiber composites will have more tendency to creep than unidirectional fiber composites [2]. Most PWB glass/epoxy composites tend to use woven E-glass fibers.

MECHANICAL SCHEME

An electronic module is connected to a motherboard through an LGA connector, as shown in Figures 16.1a and 16.1b. The connector holds 978 compliant pins in an interposer that allows vertical movement of the contacts but restrains their lateral movements. The LGA connector (Figure 16.2) has a worst-case flatness requirement of 50 microns across all pins, with a relaxation stability of an additional 13 microns. In other words, there cannot be a net relaxation of more than 13 microns of one pin relative to another due to relaxation of the PWB material.

DOI: 10.1201/9781003247005-16

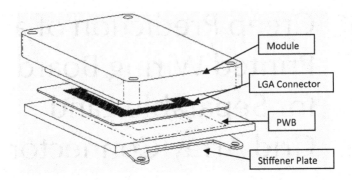

FIGURE 16.1 (a) Module mechanical stack-up.

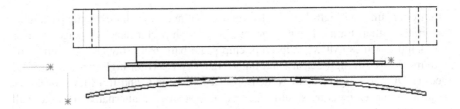

FIGURE 16.1 (b) Side profile of module stack-up. Notice curvature of the stiffener plate that flattens upon fastening.

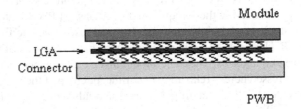

FIGURE 16.2 Module/LGA/PWB system. LGA pins behave like small springs sandwiched between the module and PWB components.

   The module/LGA assembly is attached to the PWB by the use of springs and standoffs designed to transmit 60 grams per pin on the LGA. The local stiffener plate is a low-profile flat spring designed to minimize the local warping of the PWB caused by the total clamping force of the assembly. This "flattening" of the PWB comes at the consequence of inducing a bending stress on the laminate directly underneath the module and is the driving force of the creep mechanism of the laminate. The modeled out-of-plane bending displacement of the PWB after fastening of the module is shown in Figure 16.3.

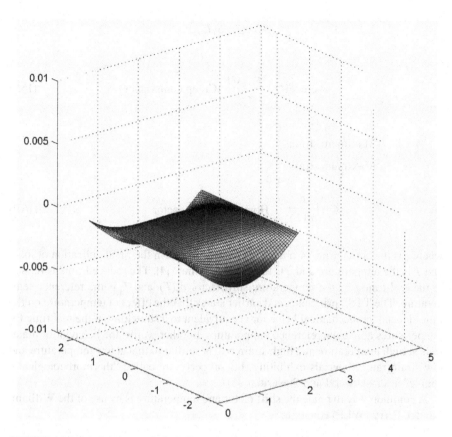

FIGURE 16.3 Modeled out-of-plane displacement of PWB after module assembly. The center of the PWB has a positive displacement while the front and back edges have a negative displacement relative to the unperturbed neutral axis of the PWB [3].

PREDICTION OF LONG-TERM CREEP

Time-Temperature Superposition (TTSP): Using the theory of linear viscoelasticity, it is possible to predict the creep over a large span of time that would otherwise be impractical to measure directly through a series of short-run creep tests at different temperatures.

Thus, a series of a handful of one-hour tests could be used to predict the lifetime creep behavior of the material by determining the creep modulus of the GETEK laminate for a given amount of stress and over a range of temperatures. The laminate is composed of 24 layers of balanced 0–90 degree woven E-glass mesh fibers, which exhibit isotropic in-plane (i.e., x-y direction) behavior. The assumption is that for small deflections, the out-of-plane displacement due to the creep will be cause by the relaxation of the in-plane modulus, and therefore the material can be treated as isotropic. For a certain class of materials, the effect due to time and temperature can be combined into a single parameter through the concept of the "time-temperature superposition principle," which implies that the following relations exist of the creep compliance:

$$S(T,t) = S(T_0, \varsigma) \text{ ,}$$

$$\text{where } S(t) = \frac{\varepsilon_{max}(t)}{\sigma_{max}} \text{ (Creep compliance)} \tag{16.1}$$

ε_{max} Maximum strain

σ_{max} Maximum stress

$$\varsigma = \frac{t}{at(T)} \quad \text{Horizontal shift factor} \tag{16.2}$$

where t is the actual time of observation measured from the first application of load, and T is the temperature and ς is the "reduced time" [4]. The reduced time is related to the real time t by the temperature shift factor $a_T(T)$, and T_0 is the reference temperature. The TTSP principle cited earlier states that the effect of temperature on the time-dependent mechanical behavior is equivalent to a stretching of the real time for temperatures above the reference temperature. In other words, the behavior of materials at high temperature and high strain rate is similar to that at low temperature and low strain rate. Materials exhibiting this property are called "thermorheologically simple" after Schwarzl and Staverman [5].

A common way to relate the shift factor and temperature is by use of the William, Landel, Ferry (WLF) equation:

$$\log_{10} a_T(t) = \frac{-k_1(T - T_0)}{k_2 + (T - T_0)} \tag{16.3}$$

where k_1 and k_2 are materials constants and T_0 is the reference temperature, which in our case is 50 °C. Two criteria must be met for a master curve to be used to predict material properties: (1) the master curve isotherms must overlap significantly and (2) the shift factor plots must be smooth curves and have no discontinuities. The procedure of TTSP for obtaining a master curve can be performed without any difficulties [6].

Steady state of creep, also called secondary creep, is successfully modeled by the power law first, which is an empirical representation suggested by Findley [7]. This yields good agreement with experimental data for steady-state creep under low stress:

$$\dot{\varepsilon} = k\sigma^p \text{ ,} \tag{16.4}$$

where $\dot{\varepsilon}$ is the steady-state creep rate, σ is the applied stress, k and p are materials constants [8]. For short loading time, the primary creep range is used and is characterized by the equation [9]:

$$\varepsilon = k\sigma^{p}t^{n}.$$
(16.5a)

Under the condition of constant stress, the equation could be further simplified to:

$$\varepsilon = Kt^{n},$$
(16.5b)

where $K = k\sigma^{p}$ and n are material constants. For values of n less than one, the slope of the strain-time curve starts vertically then decreases continuously. Thus the latter portion of a linear plot of ε versus t for any given time will look nearly like a straight line, which means that (16.5b) can be used to approximate both the primary stage and a major portion of the secondary stage without any additional terms for the second stage.

MATERIALS AND TEST SPECIMEN

The material chosen for the study was GETEK woven E-glass epoxy/phenylene resin laminate. Coupon samples, 102 mm × 12 mm, were cut out of an existing functional PWB laminate of 3.18 mm thickness in the area where the modules reside. The samples were prepared, annealed, and conditioned per ASTM D618–00 *Standard Practice for Conditioning Plastics for Testing*. To reduce error, two replicates were tested at each creep sequence.

EXPERIMENTAL SETUP

The test setup is shown in Figure 16.4 and consists of a three-point flexural bending test stand per ASTM D790, high-temperature-strain foil strain gauges, and a temperature-controlled furnace. This is a low-cost means of testing creep properties of polymers and has been employed by many researchers that test polymeric and PMC materials [10]. A dead weight is suspended above the sample material beam, and weights are added to and removed from the base as desired. Four strain gauges in a full Wheatstone bridge configuration are used to measure the maximum strain of the beam while providing compensation for thermal expansion. High temperature gauge epoxy with a T_g (glass transition temperature) of 160 °C is used to adhere the gauges. A technique of gauges attachment is used to keep the adhesive nominally at 13 microns. A convection furnace with a calibrated temperature stability of +/– 0.3 degrees Celsius is used along with a thermal couple attached to the center of the beam that is shielded from the flow of hot air blown by the fans. Specimens were tested below the glass transition temperature of the material and were subjected to a sequence of one-hour load and recovery steps. A constant stress of 7.8 MPa was applied for one hour during which strain was measured. The stress was selected based on what is believed to be worst-case induced stress on the PWB during fastening of the module. After recovery, the temperature was increased by 5 °C and the specimen was allowed to equilibrate for 10 minutes at each temperature step. Temperatures between 50 °C and 130 °C were used.

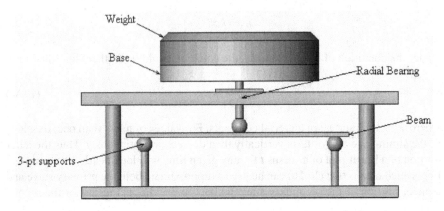

FIGURE 16.4 3-point Bending Apparatus per ASTM D790

- Material GETEK PWB laminate
  - Tg = 180 °C
  - Isothermal Tests: (50,55,60,65,70,75,80,85,90,95,100,110,115,120,125,130 °C)
  - Stress: 7.8 MPa

RESULTS

Table 1 shows the material constants used to compute the analytical WLF and Findely formulae, see Equations (3) and (6). Figure 6 shows the three-point bending results for obtaining the flexure modulus of the PCB material. Figure 7 show the results of the creep experiment performed at the indicated temperatures of 50 °C to 130 °C. At temperature below 100 °C the material exhibits relatively stable creep behavior as indicated by the small rate of change compliance vs. time. At 110 °C the creep behavior becomes apparent.

A reference temperature of 50 °C was selected for the study and was used for collapsing, by horizontal shifting, the 16 separate temperature curves used to obtain the time-temperature master curve. The master curve, shown in Figure 8, covers a span of 46 years predicted creep compliance obtained with 32 hours of total testing. The overlapping principle requires the separate curves to overlap significantly, and the shift factor curve (Figure 9) to be smooth with no discontinuities. Both requirements were satisfied by the data results.

As described previously, the PWB may not exceed 13 microns of worst-case relaxation due to creep, which translates into a creep modulus delta of 4 x10–6 (1/MPa). According to the master curve, at an ambient temperature of 50 °C and a constant stress of 7.8 MPa, this would be reached in approximately 6.5 log minutes ($10^{6.5}$ minutes), or 7.3 years. The experimental data is well matched by the theoretical representation of creep vs. stress and time as described by Findley's time-power law (Figure 10). To extend the amount of time for the PWB to relax 13 microns out-of-plane displacement, the constant bending stress of 7.8 MPa should be reduced.

TABLE 16.1

Materials Constants for Analytical Equations

William, Landel, Ferry Equation (3)		Findley Equation (5b)	
k1	−21.8	K	0.0044
k2	221.5	n	0.41
To	50 C		

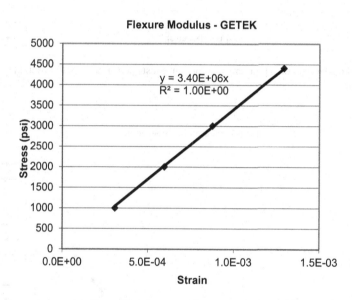

FIGURE 16.5 Flexure modulus measurements obtained from three-point bending test.

SUMMARY AND CONCLUSION

An inexpensive method of determining long term creep compliance of polymeric materials under a constant temperature and stress was investigated and developed to determine adverse effects of mechanical loading of LGA connectors onto the PWB. Creep compliance curves and a TTSP master curve was obtained from three-point bending tests. The bending samples were tested below the glass transition temperature of the laminate, and two replicate tests were performed for repeatability.

The master curve obtained from the creep tests showed that appreciable out-of-plane creep due to the constant bending load, at the specified reference temperature, would take place in the course of 7.3 years. This prediction was obtained with 32 hours of total testing at the 16 incremental temperatures. The power-law

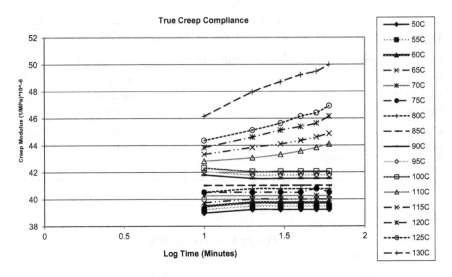

FIGURE 16.6 Result of 1-hour creep compliance tests performed for 16 different temperatures.

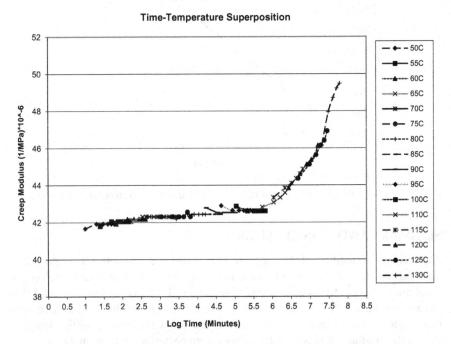

FIGURE 16.7 TTSP overlapped master curve. This curve was obtained by horizontally shifting the 16 individual creep curves obtained in Figure 16.5.

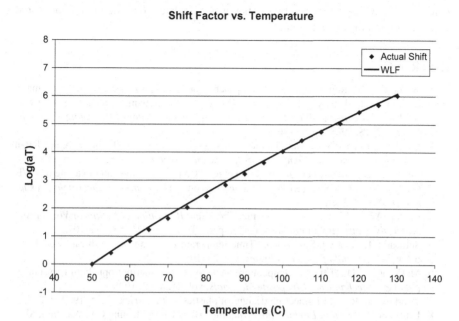

FIGURE 16.8 Shift factor plot, composed of the log of the horizontal time-shift of each of the 16 curves done in obtaining the master curve. WLF is plotted and shows good agreement.

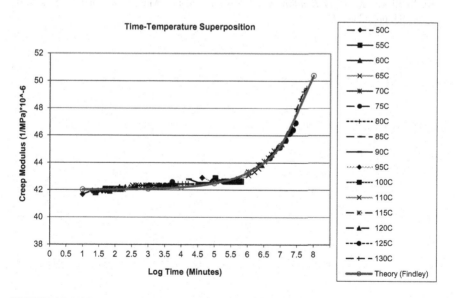

FIGURE 16.9 Master curve compared to the theoretical Findley equation..

constitutive equation proposed by Findley was calculated and fitted to the measured data and a good fit was apparent.

REFERENCES

1. Wong, C.P., Y. Rao, and J. Qu, "Creep behavior characterization of some new material for high density interconnect substrates using dynamic mechanical analyzer (DMA)," *Presented at 4th International Symposium on Advanced Packaging Materials*, March 15–18, 1998, pp. 73–76.
2. Guedes, R.M., and M.A. Vaz, "Comparison of creep behavior in UD and woven CFRP in bending," *Mechanics of Composite Materials and Structures*, vol. 8, 2001, pp. 119–134.
3. Cepeda-Rizo, J., H.Y. Yeh, and N. Teneketges, "Characterization and modeling of PWB warpage and its effect on LGA separable interconnects," *Journal of Electronic Packaging*, 2004, submitted.
4. Findely, W.N., J.S. Lai, and K. Onaran, *Creep and Relaxation of Nonlinear Viscoelastic Materials,* North Holland Publishing Company, Toronto, 1976, pp. 103–107.
5. Schwarzl, F., and A.J. Staverman, "Time-temperature dependence of linear viscoelastic behavior," *Journal of Applied Physics*, vol. 23, 1952, p. 838.
6. Maksimov, R.D., "Effect of temperature on the creep of a thermotropic liquid crystalline polymer," *Mechanics of Composite Materials*, vol. 31, no. 2, 1995.
7. Thorkildsen, R.L., "Engineering design for plastics," *SPE Series, Ch. 5*, 1964, p. 297.
8. Hult, J.A.H, *Creep in Engineering Structures*. Blaisdell Publishing Co., Waltham, MA, pp. 30–33.
9. Gates, T.S., D.R. Veazie, and L.C. Brinson, "Creep and physical aging in a polymeric composite: Comparison of tension and compression," *Journal of Composite Materials*, vol. 31, no. 24, 1997, pp. 2478–2505.
10. Mallick, P.K., *Fiber Reinforced Composites*, 2nd Ed., Marcel Dekkar Inc., New York, 1993, pp. 323–326.

17 Operational Case Studies — Mars Surface Operations

MSL REAR HAZCAM THERMAL CHARACTERIZATION

Nomenclature

AFT	= allowable flight temperature
ECAM	= engineering camera
HRS	= heat rejection system
Hazcam	= hazard avoidance camera
LMST	= local mean solar time
LST	= local solar time
MAHLI	= Mars hand lens imager
MMRTG	= multi-mission radioisotope thermoelectric generator
MSL	= Mars Science Laboratory
Navcam	= navigation camera
Sol	= Mars solar day, 24h 39m 35s

I. INTRODUCTION

The Curiosity rover (Figure 17.1) has traversed nearly 10 kilometers and has completed approximately 2.5 Earth-years of operation as of the writing of this paper. In June 23, 2014, Curiosity finished four seasons and one Martian year of operation.

The self-portrait of NASA's Curiosity Mars rover in Figure 17.1 shows the vehicle at the "Mojave" site, where its drill collected the mission's second taste of Mount Sharp. The scene combines dozens of images taken during January 2015 by the MAHLI camera at the end of the rover's robotic arm [1].

Of the total of 17 cameras on Curiosity, 12 are redundant pairs of engineering cameras and include the front and rear hazcams and the navcam, which are used in part to aid in the navigation and positioning of the rover. The hazcams pairs are mounted on the lower portion of the front and rear of the rover. These black-and-white camera pairs use visible light to produce three-dimensional (3D) terrain maps. These maps safeguard against the rover inadvertently crashing into obstacles and work in tandem with software that allows the rover to safely make some navigation choices on board.

DOI: 10.1201/9781003247005-17

FIGURE 17.1 Curiosity self-portrait at "Mojave" site on Mount Sharp.

II. DESCRIPTION

1. ENGINEERING CAMERAS

A. DESCRIPTION OF THE ENGINEERING HAZCAMS

The cameras each have a wide field of view of about 120 degrees. The rover uses pairs of hazcam (Figure 17.1) images to map out the shape of the terrain as far as 3 meters (10 feet) in front of it, in a "wedge" shape that is over 4 meters wide (13 feet) at the farthest distance. The cameras need a wide field of view because they cannot move independent of the rover; they are mounted directly to the rover body [2]. The engineering cameras, or ECAMs, are of the same design used on the twin Mars Exploration rovers, Spirit and Opportunity, that landed on Mars in 2004. The camera electronic box contains a heater resistor that warms up the electronics to above the minimum operating temperature of −55 °C. Because the detector head is thermally isolated from the electronics box, the camera electronics can be heated without significantly warming the detector head, which helps to keep thermally induced CCD dark current to a minimum [3].

B. THE REAR HAZCAM DESCRIPTION

The rear hazcams are located behind the rover on each side of the MMRTG, which powers the rover and sees temperatures near 200 °C. The hot plates (in blue in Figure 17.3) typically see temperatures of 70 °C and provide much environmental heating to the rear hazcams, so much so that they are typically 20 °C warmer than the Navcam or front hazcams.

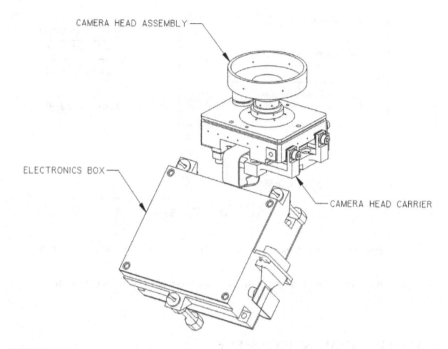

FIGURE 17.2 A schematic of the engineering camera and components.

FIGURE 17.3 There are two rear hazcams on each side of the MMRTG heat exchanger plates (blue tubing). Their proximity to the hot plates causes them to be much warmer than the other cameras.

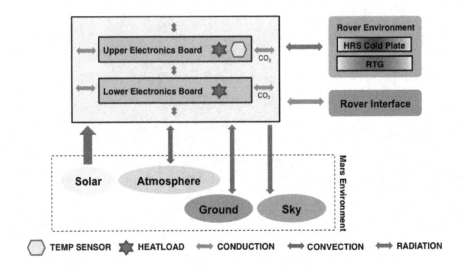

FIGURE 17.4 Description of the environmental heating loads that the rear hazcam experiences.

2. REAR HAZCAM ENVIRONMENT

One of the benefits of exposure to the hot MMRTG environment that the rear hazcams experience is that they can be operated much earlier in the morning and much later in the evening than any other cameras or actuators. A downside of the extra heating is that they are more susceptible to overheating, especially during the summer season. Figure 17.4 shows a description of the rear hazcam environment. It is worth noting that the purpose of the cold plate is to reject excess heat from inside the rover.

The cameras have two circuit boards in the electronics box that are thermally isolated from the chassis. There is thermal radiation between each board but very little thermal conduction and some thermal convection from the CO_2 gas in the Mars atmosphere. In general, the rover receives heating and cooling from solar, atmosphere, ground, and sky, as well as self-heating from usage. However, a large portion of the heating comes from thermal radiation from the MMRTG as previously described.

3. MODEL DESCRIPTION

A model was created using Thermal Desktop illustrated in Figure 17.5. The model represents the heating environment described for the rear hazcam.

MODEL CONSTRUCTION

Enclosure represented by six rectangular Thermal Desktop surfaces, with nodes combined at the interface:

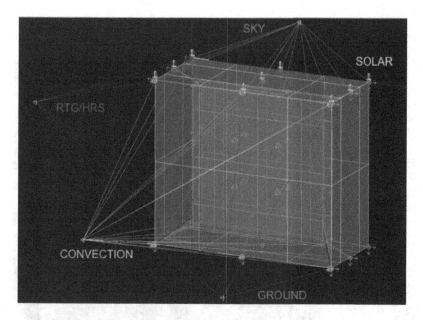

FIGURE 17.5 Thermal Desktop model.

- Material is Al 7075, 18 nodes total
  - Mounting boss features for electronics represented by diffuse nodes of equivalent thermal mass (four nodes total)
  - Each electronics board represented by a 16-node Thermal Desktop surface boards joined together by "ribbon cable"
  - Node-node contactor at each corner to represent fastener attaching boards to mounting bosses
  - Internal radiation and conduction through CO_2 was also included
  - Power-on heat load split evenly between boards
  - Flexures modeled as three node-node contactors from the enclosure to a boundary node representing the rover chassis
  - Flexures are Ti-6Al-4V
  - Flexure resistance values from ANSYS thermal FE model are 413 K/W
  - Actual fasteners used:
    — NA0070 screw (2.8 mm ø), G10 washers
  - Modeled as a parallel thermal resistance between:
    — G10 washer
      - ID = 2.4 mm, OD = 5 mm
      - Thickness = 1.6 mm on bottom washer, and 2.8 mm on top washer
      - k = 0.25 W/(m K)
    — Stainless steel bolt shank
      - k = 0.11 W/(m K)
      - Shank's conductive length = thickness of washer
      - Cross-sectional area = 6.15 mm^2

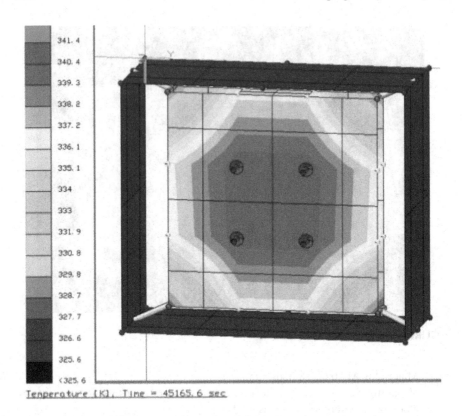

Temperature [K], Time = 45165.6 sec

FIGURE 17.6 Results of the Thermal Desktop model.

III. TESTING OF THE REAR HAZCAM ON MARS

HAZCAM THERMAL CHARACTERIZATION AND THE ECAM CALCULATOR

In order to avoid overheating the engineering cameras and to aid the planning of the camera usage, an Excel-based tool called the ECAM calculator was created that contains a data set of pre-launch rover model prediction results. The tool allows us to input time of usage and duration of camera-on times and makes a maximum temperature prediction. The calculator assumes worst-case conditions and is generally expected to produce overly conservative estimates.

In order reduce the conservative margins within the ECAM tool, an in situ thermal characterization plan for the rear hazcam was created. The plan for the rover was to incrementally increase the duration of camera use in small increments and track the temperature of the rear hazcam, in their maximum image acquiring state, all the while avoiding the rear hazcam maximum temperature limit of 50 °C. In the first run, the rear hazcam operated for 11:50 minutes, and subsequent runs added approximately 5-minute increments to the runs. A total of six runs were performed with on-times roughly ranging from 11:50 minutes to 30 minutes, as shown in Figure 17.7.

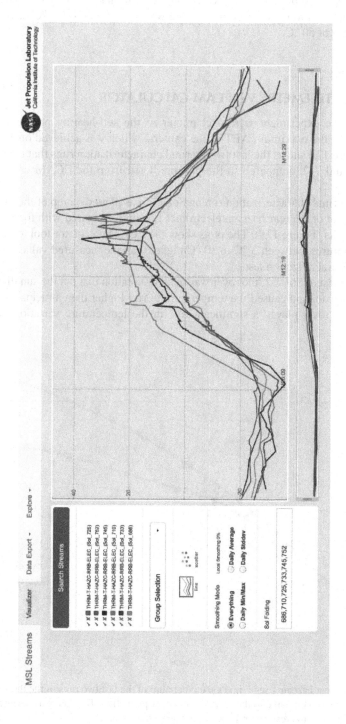

FIGURE 17.7 NASA-JPL streams query tool showing the six separate flight measurements overlapped on top of each other. The maximum peak occurred during the 25-minute camera-on run on Sol 752.

The maximum temperature occurred following 30 minutes of the camera use on Sol 752. The temperature achieved on Sol 752 was 48.5 °C, only 1.5 °C below the AFT maximum of 50 °C.

IV. RESULTS

A. HAZCAM TELEMETRY VS. ECAM CALCULATOR

The goal of the experiment was to characterize the self-heating profile in hot conditions near the maximum AFT of the camera, which was achieved on the Sol 752 run. Figure 17.8 shows the last two thermal characterization runs that occurred on Sols 745 and 752, compared to the plot predicted from the ECAM calculator tool.

The last thermal characterization run on Sol 752 was plotted on top of the diurnal temperature plot of the rear hazcam electronics box and compared with the ECAM calculator results (Figure 17.9). The plots show the ECAM calculator tool to predict hotter temperatures, between 5 °C to 10 °C higher than the measured values during the thermal characterization test.

During the test of Sol 752, the rover was in an orientation that put the sun directly on the rear hazcam and caused the temperature to rise higher than anticipated. The heading of the rover played a significant role in the temperature variation of the cameras.

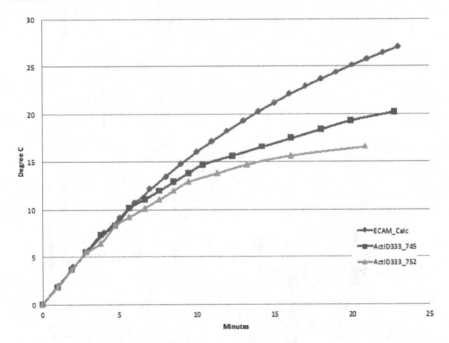

FIGURE 17.8 Rear hazcam usage time vs. temperature of the last two runs of the thermal characterization test, compared to the ECAM calculator prediction. The data was normalized to start at 0 °C so that the results could be plotted against each other.

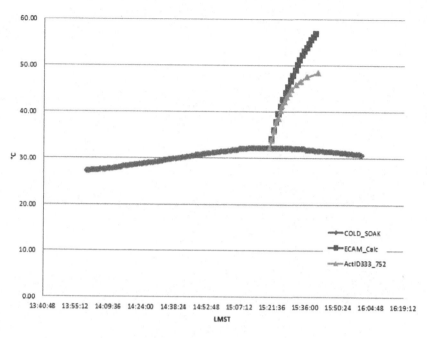

FIGURE 17.9 Rear hazcam temperature vs. usage time plotted alongside the rear hazcam "cold soak" temperature. The cold soak indicates the start temperature of the camera before powering on.

B. MODEL PREDICTION VS. TELEMETRY

The Thermal Desktop model was configured to replicate the conditions of the thermal characterization run of Sol 752, where the camera was turned with full power on at LMST 15:29 for approximately 24 minutes. Figure 17.10 compares the results of the thermal model with the telemetry during Sol 752.

V. CONCLUSION

The thermal characterization test of the rear hazcam was successfully run on the surface of Mars, which allowed us to determine how long the camera could be used without exceeding the AFT. The data from the test were plotted against and compared to the values predicted by the ECAM calculator tool, and the amount of margin that the tool overpredicted was measured. The ECAM calculator tool assumed hot case conditions and orientation relative to the sun. However, it does accommodate seasonal changes and the fact that the environment is much colder during the winter than the summer.

VI. FUTURE WORK

Future work involves creating and testing a dedicated thermal model of the rear hazcam, front hazcam, and Navcam to aid in the optimal usage of the cameras in the

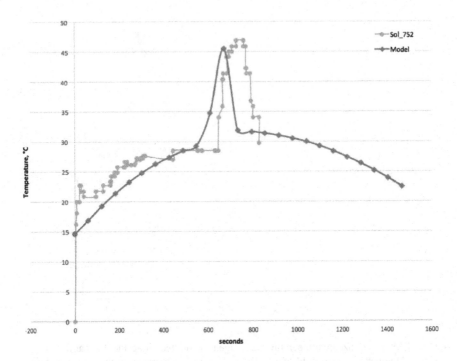

FIGURE 17.10 Rear hazcam temperature during Sol 752 thermal characterization test vs. prediction model. The model was configured with the camera on for 24 minutes, then off.

diverse environment on Mars. Focus is also being placed on creating documentation of the lessons learned that will aid in planning future Mars rover missions.

The next leg of the hazcam thermal characterization will be to run the camera for an extended period early in the morning when the danger of overheating is at a minimum. A complete profile of the cooling characteristics during the cold morning is desired so as to aid in fine-tuning our ECAM tools.

VII. ACKNOWLEDGEMENT

We like to extend our gratitude to Matthew Stumbo for spearheading the characterization testing, and to Amy Culver for planning and executing the tests.

VIII. REFERENCES

1. MSL Press Release 02.24.2015, mars.jpl.nasa.gov News Archives.
2. "Eyes and other senses," Curiosity rover camera descriptions, http://mars.nasa.gov/msl/mission/rover/eyesandother/
3. Maki, J., et al., "Mars exploration Rover engineering cameras," *Journal of Geophysical Research: Planets (1991–2012)*, vol. 108, no. E12, December 2003.
4. Cucullu, G., et al., "A curious year on mars—long-term thermal trends for mars science laboratory Rover's first Martian year," *44th International Conference on Environmental Systems*, July 13–17, 2014, Tucson, Arizona.

LASER POWER SUPPLY THERMO-STRUCTURAL ANALYSIS FOR THE MARS 2020 ROVER

INTRODUCTION

The laser power supply (LPS) is an essential part of the SHERLOC (scanning habitable environments with Raman and luminescence for organics and chemicals) instrument, slated for NASA's Mars 2020 rover (Figure 17.11) [1]. SHERLOC's LPS provides the power to the laser that will shine a tiny dot of ultraviolet laser light at a target producing a distinctive fluorescence, or glow, from molecules that contain rings of carbon atoms that offer clues to whether evidence of past life has been preserved. The laser will also induce Raman scattering, which can identify certain minerals, including ones formed from evaporation of salty water, and organic compounds. This dual use enables powerful analysis of many different compounds on the identical spot.

DESIGN CONSIDERATION: THERMAL AND ELECTRICAL ISOLATION

The LPS was designed to provide the most effective through-board thermal conduction path while providing electrical isolation to prevent corona discharge. A thermal model of the complete LPS chassis assembly, including both main printed wiring assembly (PWA) and capacitor bank PWAs, was created using SOLIDWORKS Simulation finite element code and composed of 60,267 nodes and 29,771 elements.

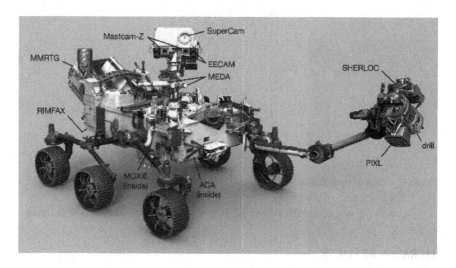

FIGURE 17.11 Mars 2020 rover showing SHERLOC on the turret assembly.

The main printed wiring board (PWB) was attached to the chassis using 10 perimeter M2 screws with a thermal resistance of 3 °C/W each. The PWB is wet-mounted to the chassis with Nusil 2946 thermal bonding adhesive around the entire perimeter. The chassis thermal attachment to the 70 °C sink is through a bracket and a flexure at each end. A total dissipation of 15.2 W was applied to the main board spread across five devices, while the capacitor banks were treated as passive. Mars atmosphere gas conduction and radiation at 70 °C were included in the model. Thermal planes were estimated across the entire main board with focus placed on careful spreading estimation underneath the high power MOSFET devices M3 and M4.

A transient analysis was performed of M3 and M4 to assure that a duty cycle of 50% of the peak power estimate was adequate. The TX2 inductor device was assumed to conduct heat away by means of an M2 screw to the chassis, while the rest of the devices conducted heat through solder pads into the main PWB.

MATERIALS

PRINTED WRING BOARD MODEL

For the PWB laminate thermal conductivity values, orthotropic values were used derived by the board stack-up for each PWB [2]. The effective thermal conductivities both in-plane of the PWB (in the XY plane of the board), and in the out-plane transverse direction (z-direction) are as follows:

$$k_{in-plane} = \frac{\sum_{1}^{N} \eta_i k_i t_i}{\sum_{1}^{N} t_i}$$

$$k_{out-plane} = \frac{\sum_{1}^{N} t_i}{\sum_{1}^{N} \frac{t_i}{\eta_i k_i}}$$

Where k_i = thermal conductivity of the ith layer,
$\quad$ t_i = thickness of ith layer,
$\quad$ η_i = % copper of the ith signal, power, or ground layer (correction factor)

THERMAL RESULTS

Figure 17.12 shows the results of the thermal analysis, with a hottest temperature of 112.4 °C measured underneath M4.

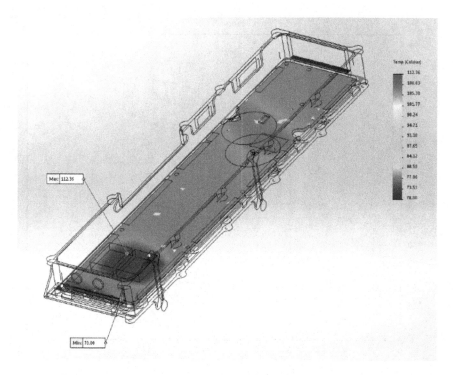

FIGURE 17.12 Steady-state maximum temperatures of MOSFETs shown as 112.36 °C for a boundary temperature of 70 °C.

TRANSIENT THERMAL ANALYSIS

A transient analysis was performed of the M3 and M4 MOSFETs to assure that a duty cycle of 50% of the peak power estimate was adequate [3]. The laser is to operate with 40 usec pulses with a total time-on of 10 seconds, and time-off of 10 seconds. The pulses are assumed to be on for 40 usec, and off for 40 usec. Essentially, there are two cycles to consider, a 10 seconds off/on main cycle, and a 40 usec on/off subcycle. It is clear that the duty cycle from an energy standpoint was 50% (e.g., the power is on only 50% of the time), but the intent is to see how the temperature varies as a function of time to entertain reducing on/off cycles as a means of reducing overall temperature. Figure 17.13 shows the 10 seconds on, 10 seconds off power cycling as well as the 40 usec pulses. Figure 17.14 shows the transient temperature across 100 seconds, and Figure 17.15 shows pulsing temperature.

STEINBERG FATIGUE ANALYSIS

A structural analysis of the LPS assembly was performed using the required environment 7.9 G$_{rms}$ random vibration spectrum for the Atlas V launch vehicle [4].

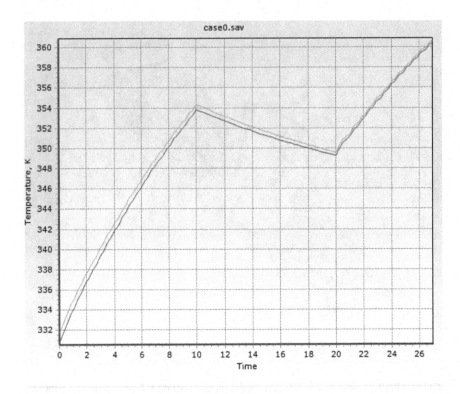

FIGURE 17.13 Thermal transient of main PWB showing min and max temperatures for 10 seconds on, 10 seconds off, and 7 seconds on.

The spectrum is applied in the worst-case direction with damping at 2%. The PWA deflection response was found to be within the guidelines of [5] for a vibration fatigue life of 20 million cycles, which is 8 hours of vibration at the fundamental PWB vibration frequency of 783 Hz (see Figure 17.16); the next mode is shown in Figure 17.17. Aluminum chassis and cover margin against yielding was 6.47. Total model mass is 1024 grams.

The lowest frequency vibration mode was the capacitor bank PWA at 783 Hz. The highest RMS von Mises stress in the chassis (Figure 17.18) was 46 MPa at the base of one of the PWA supports; peak stress was taken as the 3-sigma value, or 138 MPa. Cover stresses were less than 50% of this. PWB flexure allowance per [6] is .0107" for the main PWA and .0148" for the capacitor bank PWAs compared to the modeled 3-sigma PWB flexure of .0017" (Figure 17.19) and .0007" (Figure 17.20), respectively.

The first mode of the LPS assembly was 176 Hz (Figure 17.21) and occurred at the capacitor bank structure, which was made of polyetherimide (*ULTEM*™). The associated stress with the deflection was very small compared with the tensile strength.

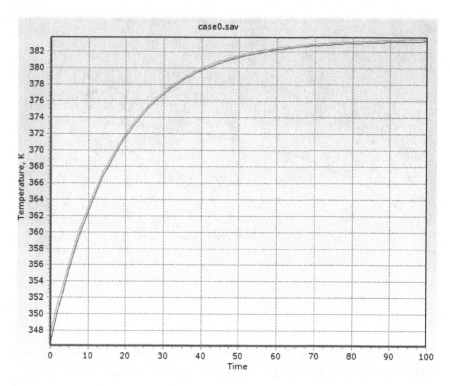

FIGURE 17.14 Thermal transient of main PWB showing min and max temperatures for continuous power across 100 seconds.

CHASSIS FASTENERS

Figure 17.22 and Table 17.1 show the results of the modeled slipping and gapping analysis results for the chassis fasteners. Table 17.2 shows the axial force value of each M2 fastener in the model. 3-sigma gapping load at the highest loaded M2 mounting fastener is 130 N. The margin was large compared to the expected M2 reduced preload of 832.5 N, or FS of +6.4.3-sigma slipping load was 61 N. The margin was large compared to the slipping capability of 166.5 N, or FS of +2.73. Positive margins are acceptable.

VENTING ANALYSIS

A venting analysis was conducted on the LPS mechanical model. The LPS was measured to have a total of 3.5 E+04 in^3 of air space inside the module. The space above and below the main PWA were assumed to be one space due to the generous amount of feedthrough area for wire routing from top to bottom of the main PWA, well below

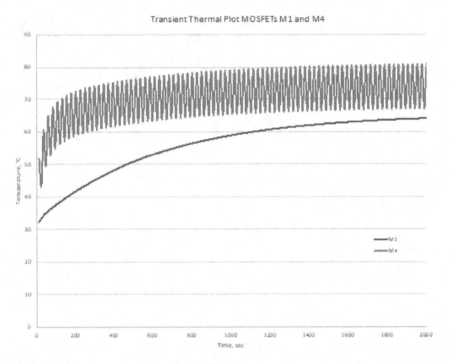

FIGURE 17.15 Thermal transient temperature showing pulsing on and off of laser power.

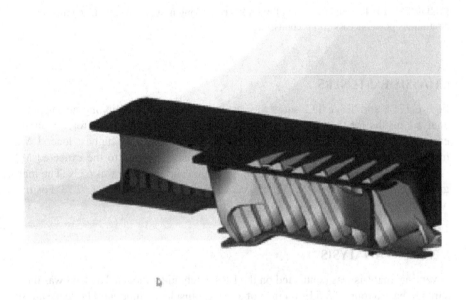

FIGURE 17.16 Fifth mode of capacitor bank PWA was 783 Hz.

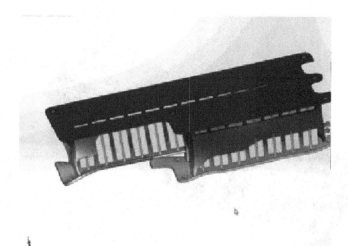

FIGURE 17.17 Sixth mode capacitor bank PWA was 1067

FIGURE 17.18 Maximum chassis stress is 4.6E+07 N/m^2.

the necessary requirement, where V is volume, and A is feedthrough area [6]. A conservative 1 psi was assumed as the pressure differential based on the Figure 17.23 guideline. The result shown in Figure 17.23 was a maximum stress at the chassis of 1.04E+08 N/m^2 for an ultimate tensile strength of 3.1E+08 N/m^2 for Al 6061-T6, rendering a factor of safety of 2.98.

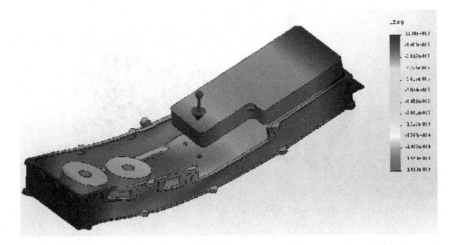

FIGURE 17.19 Main PWB modeled maximum deflection is .0017″

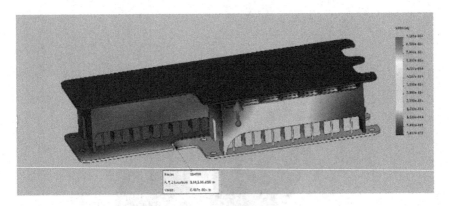

FIGURE 17.20 Cap bank PWB modeled maximum deflection is .0007″.

CONCLUSION

The analysis showed that the main PWA as well as the capacitor bank PWAs had no issues when subjected to the random vibration environment against the environment requirements. A venting analysis of the LPS assembly passed the guidelines as well.

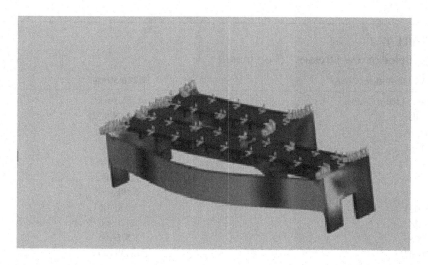

FIGURE 17.21 First mode of LPS model occurs at the capacitor bank structure and is 176 Hz. The stress associated was with the deflection was negligible.

TABLE 17.1
Gapping and Slipping Factor of Safety Results.

Fastener	Axial Force (N)	Shear 1	Shear 2	3σGapping (N) Load (N)	3σSlipping Load (N)	Reduced Load	Slipping Capability	Red. Load/ Gap. Load	Slip Cap/ Slip Load
MAX	43.33	1.6	11.2	130	61	832.5	166.5	6.4	2.73

TABLE 17.2
Modeled Screw Stresses

LPS M2 Bottom Cover	A286 Stress
Node	Value (N/m^2)
48526	1.09E+07
1297	3.98E+06
49359	1.47E+06
49090	2.03E+06
48718	2.27E+06
48813	3.10E+06
44921	1.31E+06
48850	2.99E+06
49130	1.15E+07
48385	9.56E+06
48994	4.34E+06

TABLE 17.2
Modeled Screw Stresses   (Continued)

LPS M2 Bottom Cover	A286 Stress
LPS M2 Top Cover	A286 Stress
Node	Value (N/m^2)
60276	5.89E+05
61387	3.57E+05
61342	7.12E+05
50494	2.80E+06
50501	3.29E+06
83635	3.56E+06
83734	1.42E+06
60035	8.57E+06
60007	6.43E+06
50540	3.20E+06
59911	3.89E+06

TABLE 17.3
Steinberg Fatigue Analysis Results

Steinberg PCB
Deflection Analysis

Max PCB deflection 3-sigma		Z, inches	Z, mm	PCB Length, in. B	Compt length L	PCB t h	Constant C	Placement r	PSD, G^2/HZ P
M3 & M4 LCCC, LPS Main		0.0107	0.272	9.100	0.787	0.094	2.250	1.000	0.020
LPS Caps PWBs		0.0148	0.376	9.843	1.000	0.098	1.500	1.000	0.020
Expected Main PWB Deflection (3-sigma) from FEM		0.0017	0.04318						
	Factor of Safety	6.30							

FIGURE 17.22 Chassis and M2 screws stress analysis.

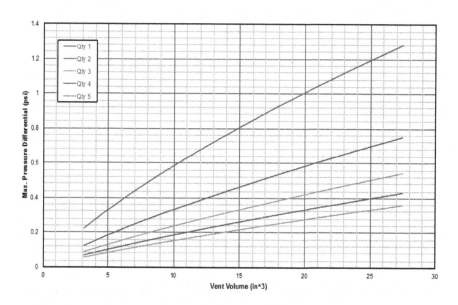

FIGURE 17.23 Vent volume vs. pressure differential.

ACKNOWLEDGEMENT

We would like to thank Rufus Simon for the design of the board electronics and Paul Eryan for leading the manufacturing effort.

The research was carried out at the Jet Propulsion Laboratory, California Institute of Technology, under a contract with the National Aeronautics and Space Administration.

REFERENCES

1. Beegle, L., et al., "Scanning habitable environments with Raman & luminescence for organics & chemicals," *IEEE Aerospace 2015*, 2015, pp. 1–11
2. Joiner, B., "Evaluation of thermal characterization techniques," *Proceedings of the IEPS Conference*, 1994, pp. 460–467.
3. Panczak, T.D., and S. Ring, *Thermal Desktop Manual*, version.6, May 17, 2017, chapter 20, pp. 65–70.
4. Dillman, R., "Planned assembly, integration, and testing of a 6M Head orbital entry vehicle," *14th International Planetary Probe Workshop*, June 12, 2017.
5. Steinberg, D., *Vibration Analysis for Electronic Equipment*, 3rd Ed., John Wiley & Son, Boca Raton, FL, 2000.
6. Kurowski, P., "Engineering analysis using SOLIDWORKS simulation 2017," SDC Publication, March 15, 2017.

18 Operational Case Studies— Dawn Asteroid Mission

INTRODUCTION

The Dawn mission was launched by NASA in September 2007 with the intent of studying the protoplanets in the asteroid belt: Vesta and Ceres [1]. Dawn entered orbit around Vesta on July 16, 2011, and completed a slightly over 1-year survey mission before embarking for Ceres in late 2012 [2, 3]. After a 2.5-year cruise, it entered orbit around Ceres on March 6, 2015 [4, 5]. Due to the failure of two of its four reaction wheels, science at Ceres was done almost entirely with the use of its reaction control system, which relied on hydrazine to achieve attitude control. In 2017, NASA announced that the planned 9-year mission would be extended until the probe's hydrazine fuel supply was depleted [6]. On November 1, 2018, NASA announced that Dawn had depleted its hydrazine, and the mission was ended. The spacecraft is currently in a derelict, but stable, orbit around Ceres [7]. Dawn is the first spacecraft to orbit two extraterrestrial bodies [8], the first spacecraft to visit either Vesta or Ceres, and the first to orbit a dwarf planet [9].

The Dawn mission was managed by NASA's Jet Propulsion Laboratory, with spacecraft components contributed by European partners from Italy, Germany, France, and the Netherlands [10]. It was the first NASA exploratory mission to use ion propulsion, which enabled it to enter and leave the orbit of two celestial bodies. Previous multi-target missions using conventional drives, such as the Voyager program, were restricted to flybys.

THERMAL DESIGN BASED ON FLIGHT PROVEN THERMAL CONTROL TECHNIQUES

Figure 18.1 shows the general construction of the Dawn spacecraft with radiators hidden, which was built around the center thrust tube assembly that contained the hydrazine and xenon tanks [10]. The two primary avionics radiators were the +Y and –Y, which contained most of the critical attitude control system (ADC) and communication systems and flight computer. The –X panel contained the visible infrared spectrometer (VIR), which required a shaded environment. The gamma ray and neutron detector (GRaND) instrument was located on the payload deck (+Z), as were the framing cameras (FC).

DOI: 10.1201/9781003247005-18

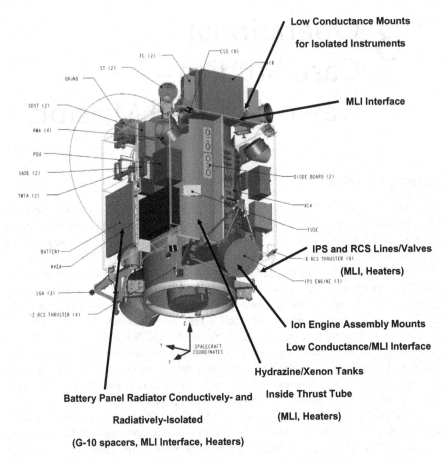

FIGURE 18.1(a) General construction of the Dawn spacecraft with hidden radiators as viewed from the –X side.

PASSIVE RADIATORS WITH OPTICAL SOLAR REFLECTORS (OSRs) REJECT HEAT DISSIPATIONS

Primary avionics and communication equipment are mounted on the –Y and +Y radiators that contain the embedded heat pipes. OSR radiators provide the high emissivity and exceptionally low solar absorptivity for the payload on the +Z instrument deck.

REDUNDANT HEATERS

All patch heaters maintain A/B redundancy and could be applied at the same time should an increase in heater power be deemed necessary. The redundancy also helps mitigate damage by micrometeor impingement.

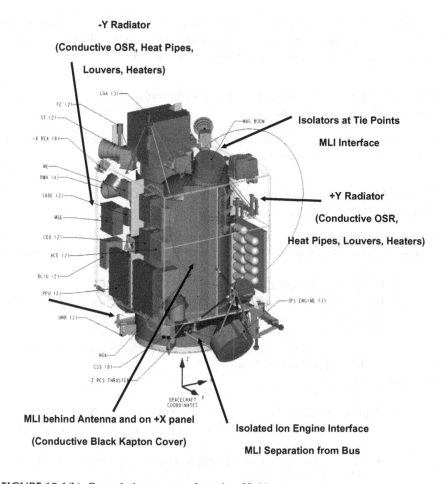

FIGURE 18.1(b) General views as seen from the +X side.

OPERATIONAL HEATERS SOFTWARE CONTROLLED

Flight software can control all heaters and can be easily reconfigured by an uplink from the ground. Set points are configurable as are duty cycles through pulse width modulation (PWM) control.

SURVIVAL HEATERS THERMOSTATICALLY CONTROLLED

Thermostatic mechanical switches are located at each radiator panel and for critical ACS systems for protection during a survival event.

MULTILAYER INSULATION (MLI) PROTECTS FROM SOLAR/SPACE ENVIRONMENT

Blanket layering consisted of 15 layers for external MLI with alternating layers of double-sided aluminized Mylar/Dacron mesh, one layer of black Kapton on the exterior, and two layers of beta cloth on the interior for micrometeor protection. The inside MLI was seven layers, with six layers alternative double-aluminized Mylar/Dacron and two layers of double-sided aluminized Kapton.

HEAT PIPES PROVIDE HIGH-EFFICIENCY HEAT SPREADING FOR LARGE DISSIPATING COMPONENTS

Twenty-eight total aluminum/ammonia heat pipes are used, 14 inches in length each, with 14 heat pipes in each Y panel in an "L shape" format.

LOUVERS MINIMIZE HEATER POWER DEMANDS DURING LOW POWER AND COLD ENVIRONMENTAL CONDITIONS

Louvers allow primarily for the saving of heater power during normal operation and are open usually only when the traveling-wave tube amplifier (TWTA) is powered on.

- Flight temperature predictions
  - Uncertainty of 5 °C carried between thermal model predictions and allowable temperatures
    - Typical GEO industry practice for long life missions
    - System performance verification testing performed generally 10 °C beyond mission predicts
- Heater sizing

  - Goal to achieve duty cycle of 75% or less at nominal 28V level
    - Heaters must maintain minimum temperatures including uncertainty at low voltage of 26 V
    - Heaters are sized to ensure current derating requirements are met at 35 V maximum voltage
  - Model uncertainty of 5 °C applies to heater sizing

The thermal subsystem design is single fault tolerant to any failure of a heater, thermostat, thermistor, or heat pipe
- Heaters
  - Primary and redundant circuits
    - Dual element or individual element heaters
- Thermistors
- Primary and redundant thermistors co-located together
- Thermostats
- Two switches in series for tolerance of a failed close thermostat
- Heat pipes
- Overlapping pipes in network to provide redundancy
- Louvers
  - Effect of single blade failure minimized by independent operation of each blade

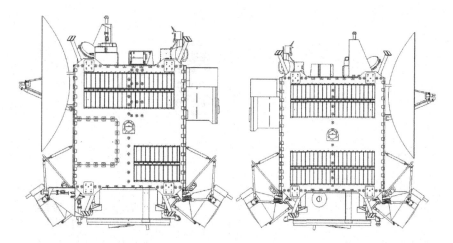

FIGURE 18.2 Louver configuration.

FIGURE 18.3 Dawn spacecraft −Y panel showing louver assembly.

OPERATION

From a thermal standpoint, the mission had most of its thermal issues and bugs early
on right after launch and during its initial 4-year cruise to Vesta [11-16]. Most of
the initial thermal problems occurred with prop-line temperatures of the electric

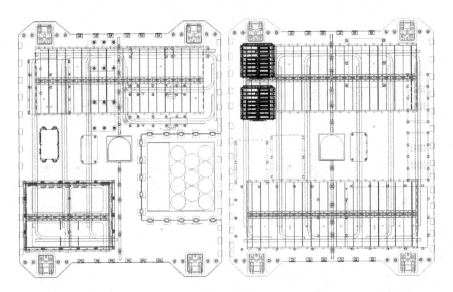

FIGURE 18.4 Heat pipe layout.

TABLE 18.1
Elements of the Spacecraft Thermal Control

Element	Location	Benefit
- Optical solar reflectors (conductive CMX Ceria-Doped Microsheet mirrors)	- ±Y, +Z radiators	- Minimizes solar influence - Proven (15 years) in deep space and GEO
- Multilayered insulation (conductive black Kapton cover)	- Partial covers on ±Y,+Z Panels - —Behind HGA and on ±X panel - Interior cylinder (propellants) - Observatory interface	- Low-risk protection from solar/ deep space environment - Extensive heritage - Ion engines assembly interface
- Aluminum heat pipes (groove type/ammonia working fluid)—constant conductance	- Embedded in ±Y Panels	- Efficient heat spreader (PPU/TW TA) - Minimizes temperature gradient and thermal distortions - Optimum mass savings for thermal control - Low-risk, proven long-term reliability
- Heaters - Operational: flight S/W control - Survival: mechanical thermostats	- Attached to interior of ±Y and -X panels - Hydrazine/xenon tanks - IPS and RCS plumbing	- Reliable protection against low temperature - Ideal for Dawn to minimize power requirements
- Louvers	- ±Y panels exterior	- Variable emittance minimizes heater demand and temperature swings - Reliable flight-proven performance
- Kapton film s/tapes, paints (black/white), and surface finishes (anodize/tiodize)	- Spacecraft interior and exterior surfaces as required	- Use of standard materials, films, and finishes with extensive flight heritage - Provides ability to tailor specific thermal performance as required

TABLE 18.2
Quantity and Mass of TCS Components

Component	Flight Quantity	Unit Mass (kg)
MLI	33 m²	12.7
OSRs	2.0 m²	1.7
Heat Pipes	42	0.12
Heaters	194	0.03
Louvers	4 at 1.4 m²	6.45
Thermistors	186	0.02
Thermostats	208	0.01

TABLE 18.3
Expected Cycles of Louvers and Thermostats

Component	Allowable Number of Cycles	Estimated Cycles	Margin (%)
Louvers	30,000	10,000	67%
Thermostats	100,000 to 1,000,000	< 90,000	tbd*

* Thermostat margin to be determined in Phase C based on specific current draws of individual circuits

propulsion system constantly triggering alarms. No thermal-related spacecraft safe-mode events happened during our tour of Vesta; the safe mode events that did occur tended to be flight software glitch related. Though Vesta had a large albedo of 0.41, arrival to the planetesimal did not reveal a significant heating of the spacecraft likely due to its lengthy distance from the sun. Most subsystems behaved as expected and the spacecraft remains healthy, though some thermal sensors required close monitoring during thrusting due to non-intuitive transient behavior, mostly in the prop lines.

Prop Line Behavior

The existing thermal model was not sufficient in predicting prop line temperature behavior. RCS lines were grouped in zones that covered too broad a temperature range, so the concern was overheating certain lines when sensors dropped too cold (see Figure 18.7). This condition frequently occurred during spacecraft maneuvers with communication to Earth and needed to be constantly monitored.

Overcompensating Heaters

Heater overcompensation tended to be a problem on certain systems including inertial relief units (IRU) and star trackers; this occurs when the heater warms the unit above the normal operating temperature. Figure 18.8 shows the set on instruments

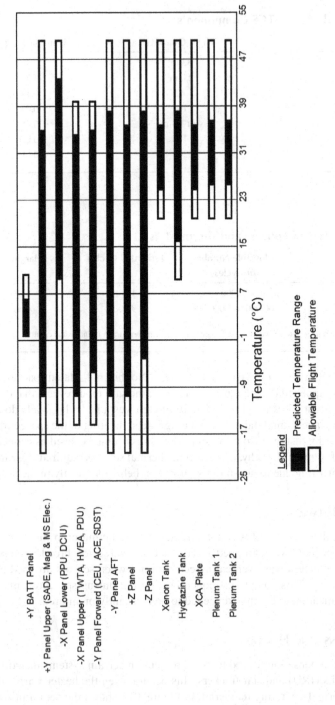

FIGURE 18.5 Predicted vs. allowable flight temperatures.

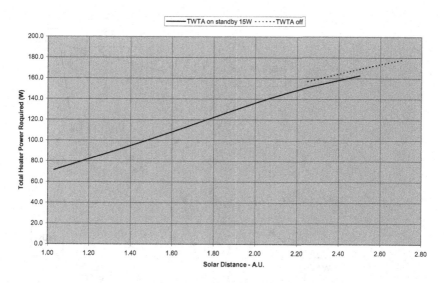

FIGURE 18.6 Total heater power required as a function of solar distance (in AU).

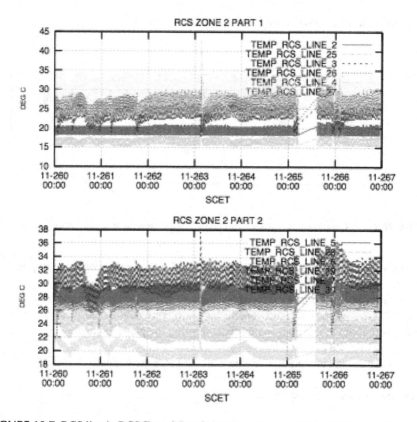

FIGURE 18.7 RCS line in RCS Zone 2 Part 2 light blue dropped below 20 °C causing a heater line in RCS Zone 2 Part 1 to heat near 45 °C, triggering the yellow alarm.

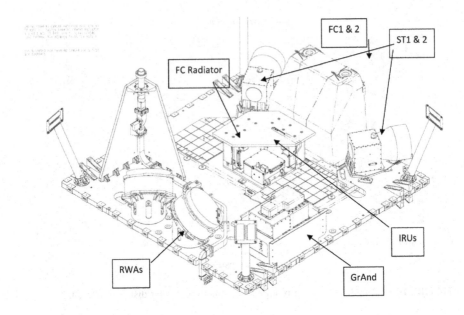

FIGURE 18.8 +Z deck set of instruments.

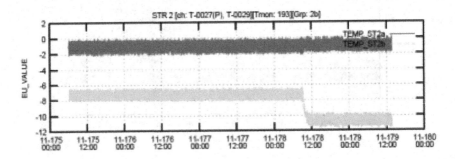

FIGURE 18.9 Star tracker 2b (green) dropped in temperature as it was powered on.

on the +Z deck, including star trackers. Figure 18.9 shows this condition as power on the star tracker 2b caused it to drop in temperature.

A summary of the general thermal issues on Dawn could be summarized as follows. These were problems in general but were consistent concerns during the science portion of the mission—as we were orbiting the planetesimals.

- No clear documentation of where all the PRTs are located.

Case: Solar Panels, IPS system.

- Lack of clear thermal system controls and spacecraft thermal software as described by one of our systems engineers: "Would be nice to see have a single knob in the software to turn up the PWM heaters."

- Thermal model too complicated; each case took one whole day to run.

REFERENCES

1. McCartney, Gretchen, Dwayne Brown, and JoAnna Wendel, "Legacy of NASA's dawn, near the end of its mission," *NASA*, September 7, 2018. Retrieved September 8, 2018.
2. "NASA's dawn spacecraft hits snag on trip to 2 asteroids," *Space.com*, August 15, 2012. Retrieved August 27, 2012.
3. "Dawn gets extra time to explore vesta," *NASA*, April 18, 2012. Archived from the original on April 21, 2012. Retrieved April 24, 2012.
4. Landau, Elizabeth, and Dwayne Brown, "NASA spacecraft becomes first to orbit a dwarf planet," *NASA*, March 6, 2015. Retrieved March 6, 2015.
5. Rayman, Marc, "Dawn journal: Ceres orbit insertion!" *Planetary Society*, March 6, 2015. Retrieved March 6, 2015.
6. Landau, Elizabeth, "Dawn mission extended at Ceres," *NASA*, October 19, 2017. Retrieved October 19, 2017.
7. Northon, Karen, "NASA's dawn mission to asteroid belt comes to end," *NASA*, November 1, 2018. Retrieved November 12,2018.
8. Rayman, Marc, "Now appearing at a dwarf planet near you: NASA's dawn mission to the asteroid belt (Speech)," Silicon Valley Astronomy Lectures. Foothill College, Los Altos, CA, April 8, 2015. Retrieved July 7, 2018.
9. Siddiqi, Asif A., *Beyond Earth: A Chronicle of Deep Space Exploration, 1958–2016* (PDF). The NASA history series (2nd Ed.), NASA History Program Office, Washington, DC, 2018, p. 2. ISBN 9781626830424. LCCN 2017059404. SP2018–4041.
10. Thomas, V.C., J.M. Makowski, G.M. Brown, et al., "The dawn spacecraft," *Space Science Revolution*, vol. 163, 2011, pp. 175–249, https://doi.org/10.1007/s11214-011-9852-2
11. "GSpace topics: Dawn," Planetary Society. Retrieved November 9, 2013.
12. "Dawn at Ceres," (PDF) (Press kit). NASA / Jet Propulsion Laboratory, March 2015.
13. Rayman, Marc, Thomas C. Fraschetti, Carol A. Raymond, and Christopher T. Russell, "Dawn: A mission in development for exploration of main belt asteroids Vesta and Ceres" (PDF), *Acta Astronautica*, vol. 58, no. 11, April 5, 2006, pp. 605–616. Bibcode:2006AcAau..58..605R. https://doi.org/10.1016/j.actaastro.2006.01.014. Retrieved April 14, 2011.
14. "Dawn," National Space Science Data Center. NASA. Retrieved November 20, 2016.
15. Chang, Kenneth, "NASA's dawn mission to the asteroid belt says good night—launched in 2007, the spacecraft discovered bright spots on Ceres and forbidding terrain on Vesta," *The New York Times*, November 1, 2018. Retrieved November 2, 2018.
16. Brown, Dwayne C., and Priscilla Vega, "NASA's dawn spacecraft begins science orbits of Vesta," *NASA*, August 1, 2011. Retrieved August 6, 2011.

19 Standards

1.0 STRUCTURAL DESIGN AND TEST REQUIREMENTS (NASA-STD-5001) [1]

TABLE 19.1

Minimum Design and Test Factors for Glass/Ceramics in Robotic Applications

Verification Approach	Loading Condition	Ultimate Design Factor	Proof Test Factor
Test	Non-pressurized	3.0	1.2
	Pressurized	3.0	2.0
Analysis Only*	Non-pressurized	5.0	N/A

*Not applicable to ceramic structures

TABLE 19.2

Minimum Design and Test Factors for Bonds in Glass/Ceramic Structures

Application	Ultimate Design Factor	Proof Test Factor
Non-pressurized	1.5	1.2
Pressurized	3.0	2.0

TABLE 19.3

Minimum Design and Test Factors for Habitable Modules, Doors, and Hatches

Pressure Load Case	Yield Design Factor	Ultimate Design Factor	Proof Test Factor
Internal pressure only	1.65	2.0	1.5
Negative pressure differential*	N/A	1.5	N/A
Negative pressure differential if verified by analysis only	N/A	2.0	N/A

DOI: 10.1201/9781003247005-19

TABLE 19.4

Minimum Design and Test Factors for Structural Softgoods

Hardware Criticality Classification	Ultimate Design Factor	Prototype Test Factor	Proof Test Factor
Loss of Life or Vehicle	4.0	4.0	1.2
All Others	2.0	2.0	1.2

TABLE 19.5

Minimum Design and Test Factors for Beryllium Structures

Yield Design Factor	Ultimate Design Factor	Proof Test Factor
1.4	1.6	1.2

2.0 DERATING STANDARDS ECSS-Q-ST-30–11-REV1 DERATING EEE COMPONENTS, PAGE 40 [2]

Derating ICs

TABLE 19.6

Derating of Parameters for Integrated Circuits

Parameters	Load Ratio or Limit	Special Conditions
Supply Voltage (Vcc)	Manufacturer recommended value ±5 90% of maximum rating	Supply voltage • Turn on transient peaks shall not exceed max rating. • Input voltage shall not exceed the supply voltage
Output Current	80%	
Junction Temperature (Tj)	110 °C or Tjmax minus40 °C (whichever is lower)	

Derating Connectors

TABLE 19.7
Derating of Family of Connectors RF Family

Parameters	Load Ratio or Limit
RF Power	75%
Working Voltage	50% of specified voltage at any altitude (pin to pin and pin to shell)
Hotspot Temperature (T_j)	30 °C below maximum rated temperature

Derating Diodes

TABLE 19.8
Derating of Parameters for Diode (Signal/Switching, Rectifier Including Schottky, Pin)

Parameters	Load Ratio or Limit
Forward Current (I_f)	75%
Reverse Voltage (V_r)	75%
Dissipated Power (P_D)	50% (only if dissipated power is defined by the manufacturer)
Junction Temperature (T_j)	110 °C or T_{jmax}—40 °C (whichever is lower)

Derating Inductors and Transformers

TABLE 19.9
Derating of Parameters for Inductors and Transformers

Parameters	Load Ratio or Limit	Special Conditions
Maximum Op. Power	50% of applied insulation test voltage	• Between winding-winding and between case
Hotspot Temperature	20 °C below max rated temp.	

Derating Resistors

TABLE 19.10
Derating of Parameters for Metal Film Precision Resistor (type RNC, except RNC90)

Parameters	Load Ratio or Limit
Voltage	80%
RMS Power	50% up to 125 °C and further decreasing to 0% at 150 °C

Note: The mentioned temperatures cited refer to case temperatures.

TABLE 19.11

Derating of Parameters for Metal Film Precision Resistor (Type RLR)

Parameters	Load Ratio or Limit
Voltage	80%
RMS Power	50% up to 70 °C and further decreasing to 0% at 125 °C

Note: The mentioned temperatures cited refer to case temperatures.

TABLE 19.12

Derating of Parameters for Foil Resistor (type RNC90)

Parameters	Load Ratio or Limit
Voltage	80%
RMS Power	50% up to 70 °C and further decreasing to 0% at 125 °C

Note: The mentioned temperatures cited refer to case temperatures.

TABLE 19.13

Derating of Parameters for Wire-Wound High-Precision Resistor (Type RBR56)

Parameters	Load Ratio or Limit
Voltage	80%
RMS Power	50% up to 115 °C and further decreasing to 0% at 130 °C

Note: The mentioned temperatures cited refer to case temperatures.

TABLE 19.14

Derating of Parameters for Chip Resistor (RM), Network Resistor

Parameters	Load Ratio or Limit
Voltage	80%
RMS Power	50% up to 85 °C and further decreasing to 0% at 125 °C

Note: The mentioned temperatures cited refer to case temperatures.

Derating Heaters

TABLE 19.15
Derating of Parameters for Heaters

Parameters	Load Ratio or Limit
Actual Rated Power (W)	50%

Derating Cables and Wires

TABLE 19.16
Derating Parameters for Wires and Cables

Parameters	Load or Ratio Limit													
Voltage						50%								
Wire size	32	30	28	26	24	22	20	18	16	14	12	10	8	6	4
Max I, Cu (A)	1,2	1,3	1,5	2,5	3,5	5	7,5	10	13	17	25	32	45	60	81
Max I, Al (A)						4	6	8	10,4	13,6	18,4	25,6	36		
Wire Temp.	Manufacturer's maximum rating T_{max} minus50 °C														
For ambient temperature of 40 °C															

For a complete list of derating spec for EEE parts, please refer to the spec.

3.0 GEVS NASA-STD-7000A[3]

The original intent for creating the General Environmental Verification Standards (GEVS) was for the use on Goddard Space Flight Center (GSFC) flight programs and projects. However, NASAs benefits by providing a set of standards to aid in mission assurance of anything that goes into space to increase the opportunity of success and decrease the chances of failure. Failed projects, even the most humanitarian in nature, can turn into rogue spacecraft that can cause harm to space access by increasing debris or forcing spacecraft to change flight paths to avoid collision. European Cooperation for Space Standardization (ECSS) similarly provides a set of universal standards made accessible to all. This standard is directed towards GSFC projects and contractors and created with the launch vehicle to be used: Atlas, Delta, and Pegasus. Given that as of this book's writing, SpaceX Falcon 9 is the most popular rocket for general transport to space, followed by India's PSLV program and US-based Rocket Lab, it is advisable to obtain the launch environment of your rocket of interest. Many of the up-and-coming launch providers are private and do not make their proprietary information publicly available, so GEVS is a good place to start.

These are guidelines for environmental verification of subsystems and components, and they describe methods for implementing those requirements, whether by test or analysis. The goal is to identify methods to ensure that the performance of hardware in the expected mission environments is satisfactory and that minimum workmanship standards have been met.

This chapter only provides a subset of the most popular sets of information on thermal/structural verification and validation of the spacecraft and its hardware. The reader should review these standards and the following for a complete coverage:

NASA Standards—The following standards provide supporting information:

a. NASA-STD 7002, Payload Test Requirements
b. NASA-STD-7001, Payload Vibroacoustic Test Criteria
c. NASA-STD-7003, Pyroshock Test Criteria
d. NASA-HDBK-7004, Force Limited Vibration Testing
e. NASA-HDBK-7005, Dynamic Environmental Criteria
f. NASA-STD-5001, Structural Design and Test Factors of Safety for Space Flight Hardware
g. NASA-STD-5002, Load Analyses of Spacecraft and Payloads
h. NASA-STD-5009, Nondestructive Evaluation Requirements for Fracture Critical Metallic Components
i. NASA-STD-5019, Fracture Control Requirements for Spaceflight Hardware

TABLE 19.17

Test Factors/Durations

Test	Prototype Qualification	Protoflight Qualification	Acceptance
Structural Loads1 Level Duration Centrifuge/Static Load[6] Sine	1.25 × Limit Load 1 minute 5 cycles @ full level per axis	1.25 × Limit Load 30 seconds 5 cycles @ full level per axis	1.0 × Limit Load 30 seconds 5 cycles @ full level per axis
Acoustics Level[2] Duration	Limit Level + 3dB 2 minutes	Limit Level + 3dB 1 minute	Limit Level 1 minute
Random Vibration Level[2] Duration	Limit Level + 3dB 2 minutes/axis	Limit Level + 3dB 1 minute/axis	Limit Level 1 minute/axis
Sine Vibration[3] Level Sweep Rate	1.25 × Limit Level 2 oct/min	1.25 × Limit Level 4 oct/min	Limit Level 4 oct/min
Mechanical Shock Actual Device Simulated	2 actuations 1.4 × Limit Level 2 × Each Axis	2 actuations 1.4 × Limit Level 1 × Each Axis	1 actuations Limit Level 1 × Each Axis
Thermal Vacuum	Max./min. predict. ± 10 °C	Max./min. predict. ± 10 °C	Max./min. predict. ± 5 °C
Thermal Cycling[4,5]	Max./min. predict. ± 25 °C	Max./min. predict. ± 25 °C	Max./min. predict. ± 20 °C
EMC & Magnetics	As Specified for Mission	Same	Same

1. If qualified by analysis only, positive margins must be shown for factors of safety of 2.0 on yield and 2.6 on ultimate. Beryllium and composite materials cannot be qualified by analysis alone. Note: Test levels for beryllium and composite structure, including metal matrix, are 1.25 × Limit Level for both qualification and acceptance testing.
2. As a minimum, the test level shall be equal to or greater than the workmanship level.
3. The sweep direction should be evaluated and chosen to minimize the risk of damage to the hardware. If a sine sweep is used to satisfy the loads or other requirements, rather than to simulate an oscillatory mission environment, a faster sweep rate may be considered, e.g., 6–8 oct/min to reduce the potential for over stress.
4. It is recommended that the number of thermal cycles and dwell times be increased by 50% for thermal cycle (ambient pressure) testing.
5. Thermal cycling testing performed as a screen for mechanical hardware with no heat generating devices may be tested to Thermal Vacuum Test factors.

TABLE 19.18
Factors of Safety

Type	Static	Sine	Random/Acoustic[4,5]
Metallic Yield	1.25[3]	1.25	1.6
Metallic Ultimate	1.4[3]	1.4	1.8
Stability Ultimate	1.4	1.4	1.8
Beryllium Yield	1.4	1.4	1.8
Beryllium Ultimate	1.6	1.6	2.0
Composite Ultimate	1.5	1.5	1.9
Bonded Inserts/Joints	1.5	1.5	1.9

1. Factors of safety for pressurized systems to be compliant with AFSPCMAN 91–710 (Range Safety).
2. Factors of safety for glass and structural glass bonds specified in NASA-STD-5001
3. If qualified by analysis only, positive margin must be shown for factors of safety of 2.0 on yield and 2.6 on ultimate.
4. Factors shown should be applied to statistically derived peak response based on RMS level. As a minimum, the peak response shall be calculated as a 3-sigma value.
5. Factors shown assume that qualification/protoflight testing is performed at acceptance level plus 3dB. If difference between acceptance and qualification levels is less than 3dB, then above factors may be applied to qualification level minus 3dB.

TABLE 19.18

Generalized Random Vibration Test Levels Components (ELV) 22.7 kg (50 lb) or Less

Frequency (Hz)	ASD Level (g²/Hz)	
	Qualification	Acceptance
20	0.026	0.013
20–50	+6 dB/oct	+6 dB/oct
50–800	0.16	0.08
800–2000	−6 dB/oct	−6 dB/oct
2000	0.026	0.013
Overall	14.1 G$_{rms}$	10.0 G$_{rms}$

The acceleration spectral density level may be reduced for components weighing more than 22 7 kg (50 lb) according to:

	Weight in kg	Weight in lb
dB reduction	= 10 log(W/22.7)	10 log(W/50)
ASD$_{(50–800\ Hz)}$	= 0.16·(22.7/W)	0.16·(50/W) for protoflight
ASD$_{(50–800\ Hz)}$	= 0.8·(22.7/W)	0.08·(50/W) for acceptance

Where W = component weight

The slopes shall be maintained at + and −6dB/oct for components weighing up to 59 kg (130 lb). Above that weight, the slopes shall be adjusted to maintain an ASD level of 0.01 g²/Hz at 20 and 2000 Hz.

For components weighing over 182 kg (400 lb), the test specification will be maintained at the level for 182 kg (400 lb).

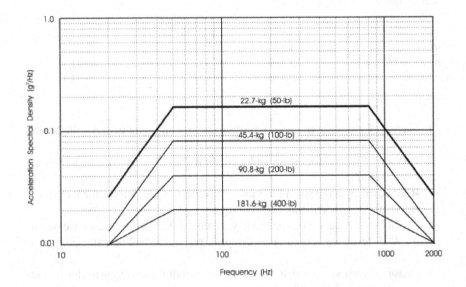

TABLE 19.19

Component Minimum Workmanship Random Vibration Test Levels 45.4 kg (100 lb) or Less

Frequency (Hz)	ASD Level (g²/Hz)
20	0.01
20–80	+3 dB/oct
80–500	0.04
500–2000	−3 dB/oct
2000	0.01
Overall	6–8 G$_{rms}$

The plateau acceleration spectral density level (ASD) may be reduced for components weighing between 45.4 and 182 kg, or 100 and 400 pounds, according to the component weight (W) up to a maximum of 6 dB as follows:

	Weight in kg	Weight in lb
dB reduction	= 10 log(W/45.4)	10 log(W/100)
ASD(plateau) level	= 0.04·(45.4/W)	0.04·(100/W)

The sloped portions of the spectrum shall be maintained at plus and minus 3 dB/oct. Therefore, the lower and upper break points, or frequencies at the ends of the plateau become:

F_L = 80 (45.4/W) [kg] F_L = frequency break point low end of plateau
   = 80 (100/W) [lb]

F_H = 500 (W/45.4) [kg] F_H = frequency break point high end of plateau
   = 500 (W/100) [lb]

The test spectrum shall not go below 0.01 g²/Hz. For components whose weight is greater than 182 kg or 400 pounds, the workmanship test spectrum is 0.01 g²/Hz from 20 to 2000 Hz with an overall level of 4.4 G$_{rms}$.

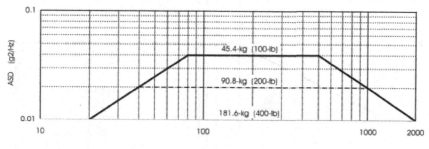

TABLE 19.20

Vacuum, Thermal, and Humidity Requirements

Requirement	Payload or Highest Practicable Level of Assembly	Subsystem Including Instruments	Unit/Component
Thermal Vacuum[1,6]	T	T	T2
Thermal Balance[1,3,6]	T and A	T, A	T, A
Temperature-Humidity[3] (Habitable Volumes)	T/A	T/A	T/A
Temperature-Humidity[4] (Transportation & Storage)	A	T/A	T/A
Leakage[5]	T	T	T

1. Applies to hardware carried in unpressurized spaces and to ELV-launched hardware.
2. Temperature cycling at ambient pressure may be substituted for thermal vacuum temperature cycling if it can be shown by a comprehensive analysis to be acceptable. This analysis must show that temperature levels and gradients are as severe in air as in a vacuum.

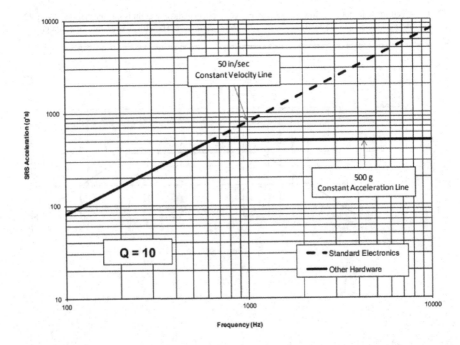

FIGURE 19.1 Shock response spectrum (SRS) for assessing component test requirements.

3. Applies to flight hardware located in pressurized area.
4. Consideration should be given to environmental control of the enclosure.
5. Hardware that passes this test at a lower level of assembly need not be retested at a higher level unless there is reason to suspect its integrity.
6. Survival/safehold testing is performed on that equipment which may experience (non-operating) temperature extremes more severe than when operating. The equipment tested is not expected to operate properly within specifications until the temperatures have returned to qualification temperatures.

T = Test required.
A = Analysis required; tests may be required to substantiate the analysis.
T/A = Test required if analysis indicates possible condensation.
T, A = Test is not required at this level of assembly if analysis verification is established for nontested elements.

Note: Card-level thermal analysis using qualification-level boundary conditions is required to ensure that derated temperature limits, for example, junction temperature limits, are not exceeded.

THERMAL MARGINS

Figure 19.2 shows operational temperature test margins. Contingency margins required by design rules are included in the development of the expected flight temperatures. Unit survival limits should be defined by the hardware limits.

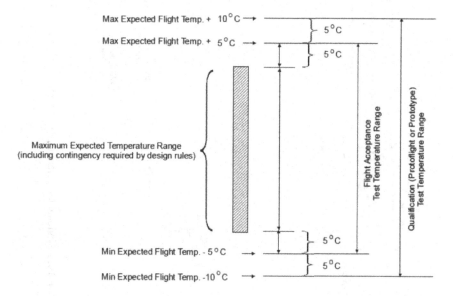

FIGURE 19.2 Qualification (protoflight or prototype) and flight acceptance thermal vacuum temperatures.

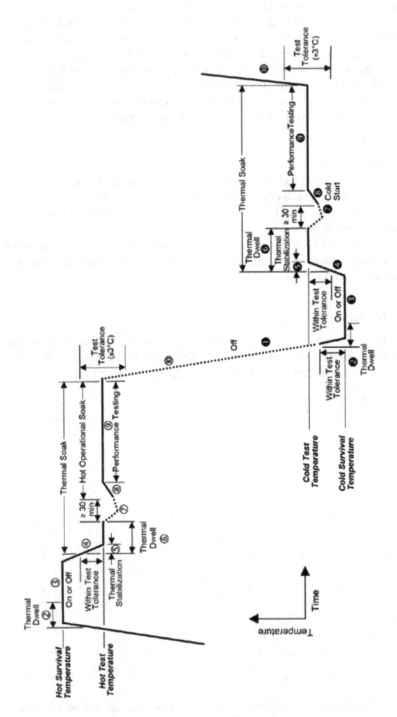

FIGURE 19.3 Notional temperature profile for the first cycle testing

4.0 SMC-160 [4]

For testing at the hot temperature on the first cycle, see Figure 19.3.

1. The environment (chamber temperature) and unit power are set to ramp the unit to its hot survival test temperature. If the hot survival temperature is an operational survival limit, the unit shall be operational (turned on either at ambient or at hot turn-on temperature) during this transition. If the hot survival temperature is a non-operational limit, the unit shall not be operating during this transition.
2. At the survival temperature, time shall be accrued to allow internal unit locations to reach survival temperature (thermal dwell).
3. The unit shall be soaked at the survival temperature.
4. The temperature shall be transitioned to the hot test temperature (e.g., qualification, proto-qualification, or acceptance). If the hot survival temperature is a non-operational limit, the unit shall be turned on at the hot turn-on temperature during the transition to the hot test temperature.
5. When the control temperature is within the test tolerance, the environment shall be adjusted to bring the control temperature to the hot test temperature.
6. Additional time should be accrued at the hot test temperature to allow internal unit locations to reach the test temperature (thermal dwell).
7. Following this dwell, the unit shall be turned off for at least 30 minutes off to allow internal temperatures to stabilize to non-operational levels. During the non-operational time, the environment may be adjusted to keep the unit temperature within the test tolerance.
8. The unit shall be turned-on and if necessary, the environment shall be adjusted to restabilize the unit at the test temperature.
9. Performance testing at the hot test temperature shall be conducted.
10. After the hot operational soak time is satisfied and performance testing is completed, the environment shall be set to ramp the unit to the cold survival temperature.

6.3.8.3.1 Option for Use of Slice/Board Testing for Acceptance Cycles

If slice/board thermal testing is performed on flight hardware, then a credit may be taken toward meeting unit-level thermal test requirements. To allow slice/ board credit toward unit level requirements, the following criteria shall be satisfied:

a. Slices/boards shall be powered on and monitored during testing.
b. All slices, boards, and cards in a unit shall be tested in the same manner. Temperature levels (average values at the same relative locations) shall envelope (hot and cold) those at the unit level.

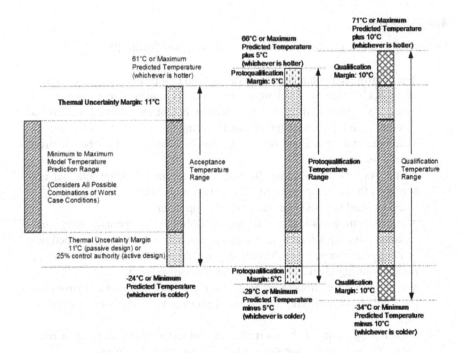

FIGURE 19.4 Unit test temperature ranges and margins

The transition rate between hot and cold shall be at an average rate of 3 °C to 5 °C per minute, and shall not be slower than 1 °C per minute.

c. Duration. The minimum number of thermal cycles (TC) shall be as shown when combined with the unit thermal vacuum (TV) test (Table 6.3–3):

Qualification:	23 TC and 4 TV cycles
Protoqualification:	16 TC and 4 TV cycles
Acceptance:	10 TC and 4 TV cycles

When an acceptance unit is insensitive to the vacuum environment (4.10.2) and only the thermal cycle test is performed (no thermal vacuum test), the minimum number of cycles is:

Acceptance:	14 TC cycles
Qualification:	$27(105/\Delta T)^{1.4}$ cycles
Protoqualification:	$20(95/\Delta T)^{1.4}$ cycles
Acceptance:	$14(85/\Delta T)^{1.4}$ cycles

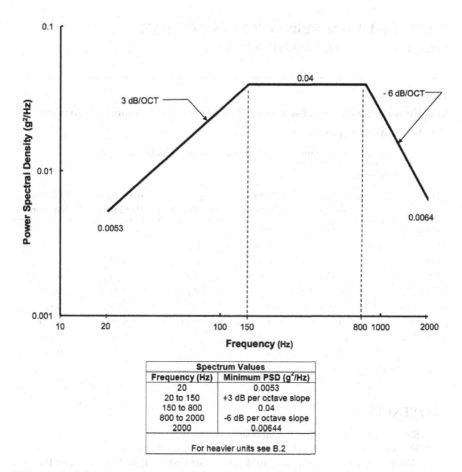

Spectrum Values	
Frequency (Hz)	Minimum PSD (g²/Hz)
20	0.0053
20 to 150	+3 dB per octave slope
150 to 800	0.04
800 to 2000	-6 dB per octave slope
2000	0.00644
For heavier units see B.2	

FIGURE 19.5 Minimum random vibration spectrum, unit acceptance test.

5.0 BOLT THERMAL RESISTANCES—SPACECRAFT THERMAL CONTROL HANDBOOK [5]

TABLE 19.21

Bolt Thermal Resistances for Use in Thermal Analyses (Source: Courtesy of The Aerospace Corporation)

Maximum Resistance Versus Bolt Size and Plate Thickness (°C/W bolt)

Size	Steel Bolt Shaft Diam (mm)	1.57 mm Aluminum	3.18 mm Aluminum	6.35 mm Aluminum	9.53 mm Aluminum
NC 4–40	2.84	12.6	–	–	–
NC 6–32	3.51	6.61	2.2	–	–
NC 8–32	4.17	4.5	1.5	0.75	–
NF 10–32	4.83	3.0	1.0	0.50	0.333
NF ¼-28	6.35	2.1	0.7	0.35	0.233
NF 5/16–24	7.92	1.5	0.5	0.25	0.167
NF 3/8–24	9.5	–	0.39	0.194	0.128
NF 7/16–20	11.1	–	–	0.16	0.106
NF ½-20	12.7	–	–	–	0.089

REFERENCES

1. Structural Design and Test Requirements, NASA-STD-5001
2. ECSS-Q-ST-30–11-Rev1, Derating EEE Components, p. 40.
3. General Environment Verification Standards for GSFC Flight Programs and Projects, GEVS NASA-STD-7000A, April 22, 2013
4. Space and Missile Systems Standards, SMC-160, September 5, 2014
5. Gilmore, D., *Bot Thermal Resistances, Spacecraft Thermal Control Handbook*, AIAA-Aerospace Corporation, Danvers, MA, 2002.

Index

A

absorptivity, 27, 28, 29, 30, 36, 38, 260
adhesive, 39, 151, 156, 157, 158, 159, 160, 161, 162, 183, 185, 209, 231, 248
adiabatic, 41, 46, 48, 50, 162
albedo, 34, 265
alumina, 75, 76, 153, 155, 156, 168
AMERSTAT, 39
ATK, 43, 46, 47, 53

B

beta cloth, 34, 36, 37, 38, 262
blackbody, 24, 25, 26, 29, 30
blackbody radiation, 24
Brownian motion, 70, 84, 87

C

choked, 99, 101, 104, 105
C junction temperature, 93, 94, 185
classical laminate theory, 125, 129, 138
commercial off-the-shelf (COTS), 1, 8, 9, 10, 119, 120, 121
condenser, 41, 42, 43, 44, 45, 46, 47, 50, 51, 52, 55, 56, 66, 67, 70, 73, 114, 119
conduction, 3, 14, 15, 16, 19, 35, 87, 95, 115, 116, 145, 146, 148, 149, 158, 240, 241, 247, 248
contact conductance, 152
contact resistance, 80, 151, 152
convection, 19, 20, 70, 71, 74, 75, 76, 79, 80, 81, 82, 94, 95, 115, 116, 145, 148, 149, 231, 240
cryogenic, 33, 42, 43, 44, 45, 46, 47, 48, 50, 52, 53
CubeSats, 3, 4, 7, 13
curiosity, 1, 3, 4, 237, 238

D

Dacron, 35, 37, 38, 39, 261, 262
decontamination, 43, 44, 45, 47, 50, 51, 52, 53
diode, 41, 42, 70, 153, 155, 156, 273

E

Earth, 34, 38, 115, 237, 265
ECOSTRESS, 37
electromagnetic interference (EMI), 38
electronic packaging (EP), 1, 5, 19
electronic packaging (EP) engineer, 1, *see also* packaging engineer

electrostatic discharge (ESD), 36, 38, 39
embossed Kapton, 35, 36
EMI, *see* electromagnetic interference (EMI)
emissivity, 26, 27, 28, 29, 30, 35, 36, 38, 146, 260
emittance, 34, 264
ESD, *see* electrostatic discharge (ESD)
Euler, 99, 101, 104, 105
evaporator, 41, 42, 43, 44, 45, 46, 47, 48, 49, 50, 51, 52, 53, 55, 65, 66, 67, 70, 73, 74, 114

F

fatigue, 151, 191, 192, 197, 198, 199, 215, 216, 217, 249, 250, 256
figure of merit, 80, 82
Fourier series, 130
freezing, 43, 54, 55, 56, 57, 58, 59, 60, 61, 62, 65

G

Grashof number, 75

H

heat pipes, 3, 15, 41, 43, 55, 56, 59, 65, 66, 68, 71, 260, 261, 262, 264, 265
heat transfer coefficient, 20, 73, 80, 84
HFE-7100, 82, 86, 90, 94, 95

I

immersion cooling, 79, 88, 91, 92, 96, 97
incompressible, 99, 101
indium tin oxide (ITO), 38
infrared (IR), 34
Interconnect and Packaging Committee (IPC), 126, 128, 131
IPC, *see* Interconnect and Packaging Committee (IPC)
IR, *see* infrared (IR)
ITO, *see* indium tin oxide (ITO)

J

Jet Propulsion Laboratory (JPL), 1, 7, 8, 9, 10, 13, 15, 37, 43, 44, 45, 46, 47, 48, 52, 53, 55, 65, 99, 110, 121, 123, 152, 165, 218, 219, 243, 246, 258, 259
JPL, *see* Jet Propulsion Laboratory (JPL)
Jupiter, 38

K

Kapitza resistance, 84
Kapton, 35, 36, 37, 38, 261, 262, 264

L

land grid array (LGA), 125, 126, 127, 128,
    131, 133, 134, 135, 136, 137. 138, 227,
    228, 233
legacy, 1, 5, 8, 9
Levy solution, 130, 131, 136, 137, 140
LGA, *see* land grid array (LGA)
liquid trap (LT), 41, 43, 45, 46
low earth orbit (LEO), 1, 7, 13
LT, *see* liquid trap (LT)

M

Mars, 1, 7, 9, 10, 20, 44, 151, 157, 197, 205, 237,
    238, 242, 245, 246, 247, 248
mass participations, 210, 212, 213, 214
methane, 44
Monte Carlo, 148, 150
Moon, 1, 34
MOSFET, 157, 158, 159, 160, 161, 162, 163, 165,
    177, 248, 249
multilayer insulation (MLI), 7, 33, 34, 35, 36, 37,
    39, 46, 260, 261, 265
Mylar, 34, 35, 36, 37, 38, 261, 262

N

nano, 70, 88
nanofluid, 43, 70, 72, 75, 76, 77, 82, 83, 84, 85,
    86, 87, 95, 96, 97
nanoparticles, 70, 75, 76, 77, 82, 83, 84, 85, 86
NASA, 1, 4, 7, 9, 35, 37, 38, 53, 55, 68, 112, 152,
    158, 237, 243, 247, 259, 275, 276, 277
Navier, 129, 131, 132, 133, 134, 135, 136,
    137, 138
Navier-Stokes, 101, 102, 105
network solver, 146
New Space, 3, 5, 7, 8, 9, 10, 13, 43
Nomex, 37, 38
Nu, *see* Nusselt number (Nu)
Nusselt number (Nu), 20, 80

P

packaging engineer, 1, 5, 23, 27, 41
phase change, 70, 148
Polyimide, 152, 167, 208
Pr, *see* Prandtl number (Pr)
Prandtl number (Pr), 20, 80, 81
pressure sensitive film, 127

printed wiring board (PWB), 125, 126, 127, 128,
    131, 132, 133, 134, 135, 136, 137, 138, 152,
    157, 158, 167, 168, 169, 171, 172, 175, 177,
    185, 188, 190, 198, 205, 207, 209, 210, 215,
    216, 217, 227, 228, 229, 231, 232, 233, 248,
    250, 251, 254, 256
pumped fluid loops, 19, 20, 79
pyrolytic graphite, 3

R

radiation, 19, 23, 24, 25, 27, 28, 29, 30, 31, 32, 71,
    115, 116, 145, 146, 147, 148, 149, 151, 158,
    167, 185, 240, 241, 248
ray trace, 152
Re, *see* Reynold's number (Re)
reflectivity, 29
reliability, 1, 2, 5, 8, 9, 38, 45, 79, 264
reservoir, 41
response, 75, 77, 191, 192, 193, 198, 205, 210,
    212, 213, 214, 216, 250, 277, 280
Reynolds, 20, 81, 84, 95, 99, 101
Reynold's number (Re), 80
Rocket Lab, 8, 10, 13, 33, 275
Rohsenow, 74, 75, 76
root mean square (RMS), 195
rover, 1, 3, 4, 7, 209, 237, 238, 240, 241, 242, 244,
    246, 247

S

Saturn, 38
second surface mirror (SSM), 36
SINDA, 146, 147, 148, 150
sintered-copper, 43
solar flux, 34
solder, 1, 79, 88, 89, 92, 93, 94, 123, 152, 158,
    185, 197, 227, 248
SOLIDWORKS, 152, 157, 165, 179, 198, 199,
    208, 210, 247
SpaceVNX, 13, 14, 15, 16
SSM, *see* second surface mirror (SSM)
Steinberg, D., 194, 197, 198, 199, 205, 208, 216

T

Teflon, 38, 39
Tesla turbine, 99, 101, 104, 107, 109, 110, 118,
    119, 121
thaw, 65, 66, 68
thermal capacity, 71
thermal conductivity, 19, 54, 62, 70, 71, 75, 76,
    79, 81, 82, 83, 84, 85, 86, 87, 94, 115, 123,
    145, 167, 168, 179
thermal interface material (TIM), 13, 16, 79, 89,
    152, 154

thermal straps, 3
TIM, *see* thermal interface material (TIM)

U

ultimate tensile strength, 208, 215, 217, 253

V

vacuum, 1, 5, 9, 10, 23, 39, 46, 47, 55, 56, 59, 65, 70, 280
variable conductance heat pipes (VCHPs), 41, 42

Velcro, 37, 38, 39, 40
venting, 38
Venus, 1, 123
vibration analysis, 197, 198, 199, 208
VPX, 13

W

water, 19, 24, 39, 41, 42, 43, 54, 55, 56, 57, 58, 59, 60, 61, 62, 63, 68, 70, 73, 75, 76, 77, 82, 247
wedgelocks, 3, 13, 14, 15
wick, *see* heat pipes

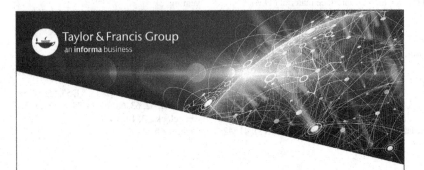

Printed in the United States
by Baker & Taylor Publisher Services